THE MUSLIM CONTRIBUTION TO MATHEMATICS

The Muslim Contribution to Mathematics

ALI ABDULLAH AL-DAFFA'

CROOM HELM LONDON

HUMANITIES PRESS
Atlantic Highlands, N.J.

Croom Helm Ltd, 2–10 St John's Road, London SW11

Reprinted 1978

British Library Cataloguing in Publication Data

Al-Daffa', Ali Abdullah
The Muslim contribution to mathematics.
1. Mathematics, Islamic—History
I. Title
510'.917'671 QA23

ISBN 0–85664–464–1

© 1977 Ali Abdullah Al-Daffa'

First published in the USA 1977 by
Humanities Press, Atlantic Highlands, N.J.

Library of Congress Cataloging in Publication Data

Al-Daffa', Ali Abdullah.
The Muslim contribution to mathematics.
Bibliography: p. 103
Includes index.
1. Mathematics, Arabic. I. Title.
QA23.D33 1977 510'.917'4927 77–3521

ISBN 0–391–00714–9

Printed and bound in Great Britain

CONTENTS

Preface

1. Introduction 9

2. Historical Setting 19

3. Arithmetic 31

4. Algebra 49

5. Trigonometry 67

6. Geometry 81

7. Conclusions 93

Bibliography 103

Index 119

PREFACE

This book presents the Muslim contribution to mathematics during the golden age of Muslim learning approximately from the seventh through to the thirteenth century. It was during this period that Muslim culture exerted powerful economic, political, and religious influence over a large part of the civilized world. The work of the Muslim scholars was by no means limited to religion, business, and government. They researched and extended the theoretical and applied science of the Greeks and Romans of an earlier era in ways that preserved and strengthened man's knowledge in these important fields.

Although the main objective of this book is to trace the history of the Muslim contribution to mathematics during the European Dark Ages, some effort is made to explain the progress of mathematical thought and its effects upon present day culture. Certain Muslim mathematicians are mentioned because of the important nature of their ideas in the evolution of modern mathematical thinking during this earlier era.

Muslim mathematicians invented the present arithmetical decimal system and formalised the fundamental operations connected with it – addition, subtraction, multiplication, division, raising to a power, and extracting the square root, and the cubic root. They also introduced the 'zero' symbol to Western culture. This simplified considerably the entire arithmetical system and its fundamental operations; it is no exaggeration if it is said that this specific invention marks one of the significant turning points in the development of Mathematics.

During the ninth century Al-Khwarizmi, the founder of algebra, transformed the concept of a number from its earlier arithmetic character as a fixed quantity into that of a variable element in an equation. He also found a method to solve general equations of the first and second degree in one unknown by both algebraic and geometric means. In addition to devising methods for solving simultaneous linear and even some types of quadratic systems, Muslim mathematicians laid the foundations for solving third and fourth degree equations.

In the domain of trigonometry, the theory of the functions 'sine', 'cosine', and 'tangent' was developed by Muslim scholars of that time, with Mohammed ibn Jabir Al-Battani being considered 'the father' in this field of mathematics. The Muslim scholars worked diligently in the development of Plane and Spherical Trigonometry and proceeded to establish those subjects as sciences independent of Astronomy.

In the field of geometry, Muslim mathematicians added considerably to the work of the Alexandrians and the Greeks by translating into Arabic and interpreting Greek contributions, solving various problems, and making comments. Abu Ali Al-Hasan ibn Al-Haitham and Thabit ibn Qurra were the foremost workers in these projects. They kept science alive when other people were ignoring it. Europe received its Greek geometry through the Muslims.

To summarize the achievements of the Muslim mathematicians: they generalized the concept of numbers beyond what was known to the Greeks. They developed and systematized the science of Algebra and preserved its link with Geometry. The work of the Greeks in Plane and Solid Geometry was continued. Finally, the Muslims developed Trigonometry, both plane and solid, producing accurate tables for the trigonometric functions and discovering many trigonometric identities.

Ali A. Al-Daffa'

1 INTRODUCTION

Muslims directed their attention to intellectual activities during the early days of Islam, approximately 700 AD, turning first to the practical sciences, such as mathematics and astronomy.

There was a religious basis to the Muslims' need for mathematics and astronomy. From geometric means they could find the direction to Mecca, towards which they turn daily in their prayers; arithmetic and algebra were needed to calculate inheritances and to count days and years. From astronomy, Muslims could determine the beginning of Ramadan, the month of fasting, and other great holy days.

However, Muslims did not confine the application of the sciences which they developed to the needs of religion but extended them in many directions for the benefit of mankind. When Muslims turned to the field of mathematics, which is often referred to as the 'mirror of civilization,' they were following an arduous road to cultural development.

Muslims had a great advantage in the Middle Ages because 'the Koran encouraged them to study the sky and the earth to find proofs to their faith. The Prophet Mohammed Himself had besought his disciples to seek knowledge from the cradle to the grave, no matter if their search took them as far afield as China, for "he who travels in search of knowledge, travels along God's path of Paradise." '[1]

In Western countries mathematics was at a low ebb during the Dark Ages, but the great triumphs of Greek intellect were not lost to mankind. Muslims, under the influence of a great religious impulse had become a very rapidly expanding power. [2]

From 600 to 1200 AD, the Muslim Empire stretched from India to Spain, Baghdad and Cordova being the centers for the reigning Caliphs.[3] The ninth and tenth centuries may be regarded as the golden age of Muslim mathematicians to whom the world owes a great debt for preserving and expanding the classics in Greek mathematics otherwise lost. Europe owes its Renaissance to this golden age. A vast field for further investigation in this connection is left for the attention of future scholars.[4]

Particular attention should be given to the spectacular rise and decline of the Muslim State that occurred during the period of Europe's Dark Ages. Within a decade following Prophet Mohammed's flight from Mecca to Medina in 622 AD, the scattered and disunited tribes of the Arabian peninsula (modern Saudi Arabia) were consolidated by a strong religious fervor into a powerful nation. After Prophet Mohammed's death, the caliphs not only governed wisely and well but many became patrons of learning and invited distinguished scholars to their courts. Numerous Hindu and Greek words in astronomy and mathematics were translated into the Arabic language and were saved. Later European scholars were able to retranslate them into Latin and other languages.[5]

About 800 AD, the city of Baghdad became a center of learning under the Muslim caliphate. The great caliph Al-Ma'mun, who was a scholar, a philosopher, and a theologian, established his famous 'House of Wisdom' (Bait Al-hikma), a combination of library, academy, and translation institution which proved to be the most important educational establishment since the foundation of the Alexandria Museum in the first half of the third century BC.[6]

Caliph Al-Ma'mun set scholars to work translating all the great Greek texts into Arabic. Thus the works of Ptolemy, Euclid, Aristotle, and many others were eventually circulated from Baghdad to Islamic universities as far away as Sicily and Spain. Through these Spanish universities established by Muslims, scientific knowledge was transferred to Europe during the Dark Ages.[7]

It was this period of the Dark Ages that may be called the Muslim Age in the history of mathematics. Princes, religious groups, and rich patrons rivalled with one another in commissioning the translation of ancient, and the writings of new, scientific masterpieces for which purpose they employed Muslim, Christian, Jewish, and even Zoroastrian scholars. All these men, though differing in religion and race, had one thing in common. They all wrote in Arabic,[8] the Muslim language.

> Abul Rihan Mohammed ibn Ahmed Al-Biruni declared that Arabic was the language of science and that he preferred Arabic curses to Persian praise. Adnan has explained that most of the textbooks used in the Turkish universities were written in Arabic, which remained the language of science in Turkey

until the 18th century.[9]

In his monumental *Life of Science,* Professor George Sarton of Harvard University has stated:

> I must insist on the fact that, though a major part of the activity of Arabic writing scholars consisted in the translation of Greek works and their assimilation, they did far more than that. They did not simply transmit ancient knowledge, they created a new one . . . However, a few Greeks had reached, almost suddenly, extraordinary heights. That is what we call the Greek miracle. But one might speak also, though in difference sense, of an Arabic miracle. The creation of a new civilization of international and encyclopaedic magnitude within less than two centuries is something that we can describe, but not completely explain.[10]

It is with the hope of making a contribution to such knowledge that the present book was begun. As a first step, it seemed advisable to coordinate Muslim contributions to mathematics of this period and interpret the findings to provide a basis for future writing. Professor George A. Miller of the University of Illinois wrote:

> The history of mathematics is the only one of the sciences to possess a considerable body of perfect and inspiring results which were proved 2000 years ago by the same thought processes as are used today. This history is therefore useful for directing attention to the permanent value of scientific achievements and to the great intellectual heritage which these achievements present to the world.[11]

Scope of Muslim Contribution

The authoritative work on the history of Muslim mathematics has yet to be written. The author's objective is to present a brief historical development of Muslim contributions to mathematics

Fig. 1.1: The rise of Muslim Science

CENTURY	DESCRIPTION	RESULT
Seventh	Birth of the Prophet Mohammed about 570 AD	Beginning of Islam 622 AD
Eighth and Ninth	The impetus	Period of consolidation of Muslim people
Tenth	The Muslim Age	The rise of Muslim scholarship
Eleventh	The Golden Age of Muslim thought	Encouragement of Muslim experimental and theoretical sciences
Twelfth and Thirteenth	A turning point	Decline of the Muslim state and rise of western culture

and the preservation of Greek and Hindu mathematics. According to F.W. Kokomoor '. . . there was probably not a single important work of the Golden Age of Greece that had not been translated and mastered by the Arabs.'[12]

The Muslims developed a vast knowledge of their own in the field of mathematics. They accomplished some scholarly work which carried mathematics beyond the limits attained by the Greeks. In particular, this was true in the areas of algebra and trigonometry.[13]

Muslims not only contributed to mathematics but also to astronomy, medicine, geography, chemistry, pharmacology, and agriculture. The author has limited the present work to the mathematical attainments of Muslims.

Some Muslim Mathematicians

From the eighth to the thirteenth centuries, Muslims shared a common culture using Arabic as its vehicle. The present study is limited to the work of those mathematicians whose contributions were in the Arabic language.

Abu Bekr Mohammed ibn Al-Hosain Al-Karkhi was born in Karkh, a suburb of Baghdad, and died in the decade 1019-29. A Muslim mathematician at Baghdad, he wrote on arithmetic, algebra, and geometry.[14]

Living during part of the ninth century (780-850) was Mohammed ibn Musa Al-Khwarizmi, the 'father of algebra.' A Muslim mathematician and astronomer, he was summoned to Baghdad by Al-Ma'mun and appointed court astronomer. From the title of his work, *Hisab al-Jabr w-al-Mugabalah* (Book of the Calculations of Restoration and Reduction), algebra (al-jabr) derived its name. Al-Khawarizmi translated certain Greek works.[15]

Mohammed ibn Jabir ibn Sinan Abu Abdulah Al-Battani was born in Battan, Mesopotamia in 850 and died in Damascus in 929. An Arab prince and governor of Syria, he is considered the greatest Muslim astronomer and mathematician. Al-Battani improved trigonometry and computed the first table of cotangents.[16]

Waijan ibn Rustam Abu Sahl Al-Kuhi lived about 985 as a Muslim astronomer and geometrician at Baghdad. He devoted his attention to Archimedian and Apollonian problems which led to equations higher than the second degree.[17]

Born at Harran in Mesopotamia in 833, Thabit ibn Qurra died at Baghdad in 902. As a mathematician and astronomer, he translated the works of Greek mathematicians and wrote on the theory of numbers.[18]

Muslim mathematicians recognized that culture is an intelligent interest in the past, present, and future achievements of men. As it was in the days of ancient Egypt and Rome, mathematics was a tool for the solution of everyday problems.[19]

According to Professor Eric Temple Bell of California Institute of Technology:

> In all historic times all civilized peoples have striven towards mathematics. The prehistoric origins are as irrecoverable as those of language and art, and even the civilized beginnings can only be conjectured from the behaviour of primitive people today. Whatever its source, mathematics has come down to the present by two main streams of number and form. The first carried along arithmetic and algebra; and the second, geometry.[20]

The Muslim Antecedents of the European Renaissance

The purpose of this book is to trace the history of Muslim mathematicians with particular attention to those contributions which have not received sufficient attention and recognition. This study could be useful to students of mathematics, especially to those of Islamic background. The author's main objective is not simply to record isolated discoveries, but rather to explain the progress of mathematical thought.

Muslims have provided considerable knowledge about mathematics, yet many Americans and Europeans do not acknowledge from what storehouse the Christian World acquired the tools without which Western civilization could not have reached its present level. While Christianity provided the West with spiritual and moral foundations, it was classical Greece that gave the West their logic, their mathematics, and many of their basic scientific techniques. For many centuries, the European mathematician arrived at his rational deductions by the method of Aristotelian logic, a logic that he imbibed from childhood, even though one is hardly aware of its existence as a factor in mental processes. Equally, man's knowledge of the most elementary truth of the various sciences has derived from such men as Pythagoras, Archimedes, and Euclid.[21]

The truly surprising thing is that despite Western dependence upon Greek knowledge, Europe failed to produce a continuous tradition stemming from it. After the fall of the Roman Empire, for about half a millennium, outside the church, Europe remained generally ignorant of her Greek patrimony. There were considerable numbers of Nestorian, Jacobites, and other monks intimately acquainted with Greek knowledge, but they were scattered among Muslims, particularly in Mesopotamia. In Paris, Oxford, and Rome, even the 'educated' citizen knew nothing of Euclid beyond his name, and had no idea of the vast amount of knowledge that the Greeks had bequeathed to the West. Yet the gulf was bridged. The bridge-builders were the Muslims.[22]

Professor Dirk J. Struik of Massachusetts Institute of Technology writes:

The Greek tradition was cultivated by a school of Arabic

scholars who faithfully translated the Greek classics into Arabic – Apollonius, Archimedes, Euclid, Ptolemy, and others. The general acceptance of the name 'Almagest' for Ptolemy's 'Great Collection' shows the influence of the Arabic translation upon the West. This copying and translating has preserved many a Greek classic which otherwise could have been lost. There was a natural tendency to stress the computational and practical side of Greek mathematics at the cost of its theoretical side. Arabic astronomists were particularly interested in trigonometry.[23]

In his book *A Guide to the History of Science*, Sarton has asserted that in explaining Western culture, one could almost omit Hindu and Chinese developments in mathematics, but to leave out Muslim developments would spoil the whole story and make it unintelligible. Muslims were standing on the shoulders of their Greek forerunners just as the Americans were standing on the shoulders of the Europeans. Arabic was the international language of mathematics to such a degree that had never before been equalled by another language except Greek and Latin. The Arabic culture was, and to some extent still is, a bridge, the main bridge between East and West. Latin culture was Western; Chinese culture, Eastern; but Arabic culture was both. It was stretched out between the Christianism of the West and the Buddhism of the East and touched both.[24]

Robert Briffault summarizes the matter in *The Making of Humanity* by stating that 'mathematics is the most momentous contribution of Arab civilization to the modern growth in which the decisive influence of Islamic culture is not traceable. Nowhere is it so clear . . . as in the genesis of that power which constitutes the permanent distinctive force of the modern world, and the supreme source of its victory – natural science and the scientific spirit.'[25]

Summary

It has become fasionable in some circles to claim that religion was an impediment to mathematical advance in Europe. The mathematical accomplishments of Muslims proved different for Islam. What, in the West, often hindered mathematical progress was the narrowly

Fig. 1.2: Muslim Learning and the Western Lines of Influence

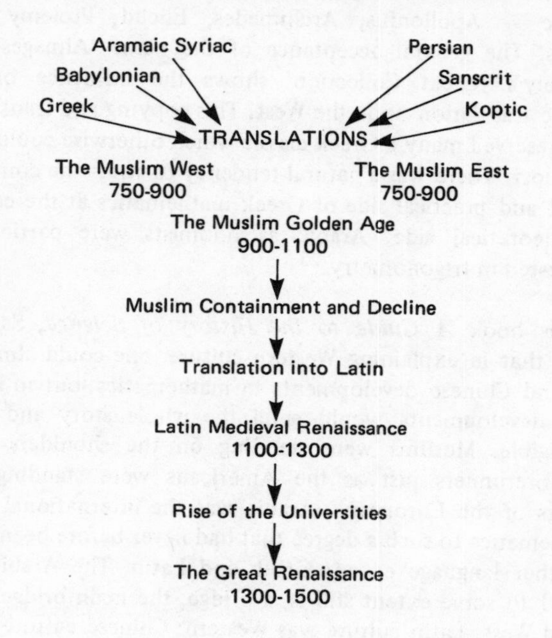

dogmatic interpretation of religion by church authorities, of which the persecution of Galileo is an example. Most of the mathematical discoveries of the Muslims came about because of Islam. Likewise, it was Islam that induced Muslim mathematicians not to limit themselves to one particular field but to become universalists.[26]

Notes

1. Rene Taton, *History of Science* (Ancient and Medieval Science from the Beginnings to 1450) (New York, Basic Books, 1963), pp. 385-6.
2. J. W. N. Sullivan, *The History of Mathematics in Europe* (London, Oxford University Press, 1925), p. 13.
3. Florian Cajori, *A History of Mathematics* (London, Macmillan, 1924), p. 99.
4. David Eugene Smith, *History of Mathematics* (New York, Ginn and Company, 1923), Vol. 1, p. 177.

5. Howard Eves, *An Introduction to the Foundations and Fundamental Concepts of Mathematics* (New York, Rinehart and Company, 1958), p. 44.
6. Carl B. Boyer, *A History of Mathematics* (New York, John Wiley and Sons, 1968), p. 251.
7. Stephen F. Mason, *A History of the Sciences* (New York, Collier Books, 1962), p. 96.
8. Rene Taton, *History of Science* (Science in the Nineteenth Century) (New York, Basic Books, Inc, 1965), pp. 572-3.
9. Ibid.
10. George Sarton, *The Life of Science* (Essays in the History of Civilization) (New York, Henry Schumann, 1948), pp. 150-1.
11. George A. Miller, *Historical Introduction to Mathematical Literature* (New York, The Macmillan Company, 1916), pp. 17-18.
12. F. W. Kokomoor, 'The Status of Mathematics in India and Arabia during the "Dark Ages" of Europe,' *The Mathematics Teacher*, Vol. 29 (January 1936), p. 229.
13. Edward S. Atiyah, *The Arabs* (The Origins, Present Conditions, and Prospects of the Arab World) (Edinburgh, R. and R. Clark, 1958), pp. 55-6.
14. Stephan and Nandy Ronart, *Concise Encyclopaedia of Arabic Civilization* (The Arab East) (New York, Frederick A. Praeger, 1960), p. 284.
15. Ibid., p. 295.
16. Ibid., p. 86.
17. George Sarton, *Introduction to the History of Science* (From Homer to Omar Khayyam) (Baltimore, The Williams and Wilkins Company, 1927), Vol. I, p. 665.
18. Carl Fink, *A Brief History of Mathematics* (Chicago, The Open Court Publishing Company, 1900), p. 320.
19. W. W. Rankin, 'The Cultural Value of Mathematics,' *The Mathematics Teacher*, Vol. XXII (April 1929), p. 215.
20. Eric Temple Bell, *The Development of Mathematics* (New York, McGraw-Hill Book Company, 1940), p. 3.
21. Rom Landau, *Arab Contribution to Civilization* (San Francisco, The American Academy of Asian Studies, 1958), p. 7.
22. Ibid., pp. 7-8.
23. Dirk J. Struik, *A Concise History of Mathematics* (New York, Dover Publications, 1948), Vol. 1, p. 92.
24. George Sarton, *A Guide to the History of Science* (Waltham, Mass., The Chronica Botanica Company, 1952), pp. 28-30.
25. Robert Briffault, *The Making of Humanity* (New York, The Macmillan Company, 1930), pp. 139-40.
26. Rom Landau, 'Arabist on the Cultural Heritage of the Arab World,' *The Arab World*, Vol. VI, Number 9 (September/October 1960), p. 13.

2 HISTORICAL SETTING

To those who delight to study man's pastoral simplicity or to moralize about the destiny of nations, the history of Arabia cannot fail to be of interest. From time immemorial Arabia has been celebrated for its cultural achievements and has distinguished itself as the home of liberty and independence by being the only land in all antiquity that has never bowed to the yoke of foreign conquerors.[1]

By the fifteenth century, the religious culture, which began in the early decades of the seventh century as purely Arabic in origin, had developed into a unique system of belief. It was established through the blending of elements from the civilizations of the highly advanced peoples who had been conquered. There were elements from Persian and Indian sources and from the ancient Roman and Greek civilizations in Spain and Sicily and traces of influences from the preceding Roman and Visigothic Germanic cultures.[2]

The Beginning of Islam

Mohammed, born in Mecca in 570 AD, was the posthumous son of Abdallah, an esteemed merchant of moderate means, and of Aminah. The death of His mother left the child an orphan when He was barely six years old.[4]

The boy first grew up under the care of His paternal grandfather, Abd Al-Muttalib, and was later entrusted to an uncle, Abu Talib. The latter was the father of Ali who was to become one of Mohammed's most faithful companions, His son-in-law, and finally His fourth successor as Caliph.[5] What is certain among the facts of the youthful life of Mohammed prior to His calling to prophethood was His marriage to the rich widow Khadijah. It was for her that He had traveled and traded. His contacts had been with Christian peoples of the desert and with the Hanifs.[6]

The Prophet Mohammed initiated the preaching of Islam which professes monotheism as expounded in the most ancient Koranic verses. It was the same God who transmitted the subsequent revelations. In the beginning Mohammed had great difficulty in convincing the peoples of Mecca of His divine vocation, and He was forced to flee to Medina, barely escaping with His life.[7]

The situation changed rapidly, however, and the most valiant and influential Quraysh members of His tribe were converted by Mohammed and went to His support.[8] These included His most faithful companions of the first hour, namely, Ali, Abu Bakr, and Omar, as well as the future great captains, namely, Khalid ibn Al-Walid (who in the future was to become the 'Sword of Allah'), Amr ibn Al-'As, and Saad ibn Abi Waqqas.[9]

From Medina, where Mohammed had returned after His Meccan triumph,[10] he reappeared once more in His native city during the tenth year after Hegira (the night flight to Medina) in order to conduct personally the Muslim pilgrimage. The year before, He had been presented by the faithful Abu Bakr. This was the 'pilgrimage of farewell,' where, with solemn words of the last revelation, the Prophet declared His mission accomplished and proclaimed that the grace of Allah had now entirely descended on His people with Islam.[11] Shortly thereafter, Prophet Mohammed died; He had not designated anyone to succeed Him but had left it to the discretion of His own people to determine who would be the next Caliph.[12]

The Caliphate

For a short period of time, the disappearance of Mohammed plunged the Medinese community, which continued to form the directing nucleus of young Islam, into a crisis. It was necessary to provide a successor who, without being able to inherit the untransmittable religious prerogatives of the Prophet, would be His successor as the political leader of the society of believers. The vital organization which Mohammed had formed under the sign of Islamic faith would not die with Him.[13] After stormy periods, due to competing individuals, Omar's energetic intervention prevailed. At the tumultous council held in the headquarters of the Banu Saidah in Medina, Omar endorsed Abu Bakr as Khalifat Rasul Allah.[14] Abu Bakr, the oldest and most faithful Meccan 'companion' of the Prophet, rose to the leadership (632-5 AD) of the Muslim community. He was a man of balance, honesty, and loyalty, as indicated by the epithet *as-siddiq* ('the veracious') which tradition had bestowed upon him. Though humble in spirit, he was inflexible in keeping custody over the precious legacy that had been entrusted to him.[15]

In 633 the Arabs penetrated into Palestine, Transjordania, and entered Syria in 644 on the flank of the Byzantines, who were engaged in a confrontation with the Arabians.[16] Abu Bakr designated Omar Al-Khattab as his successor when he died. Omar ruled the Islamic Empire from 634 to 644, the crucial decade of its formation.[17] By his impressive personal traits, he was considered by later Muslim generations as one of the most eminent of the four 'well guided' (this is how the Arab *rashidun* is usually rendered) Caliphs who formed the golden age of Islam.[18]

During the time of Omar ibn Al-Khattab, the Muslims penetrated Egypt and from there they moved toward North Africa.[19] Omar continued the expansion, not only by conquering the Roman dominions outside Europe, but also the entire domain of the Sassanian Empire.[20] While on his deathbed, Omar entrusted the nomination of a successor to a council (Shura) of six eminent Muslims. They had all been companions of Mohammed: Ali, Talhah, Zubayr, Abd ar-Rahman ibn Awf, Saad ibn Abi Waggas, and Othman ibn Affin. The last became Caliph (644-56). He was a member of the aristocratic Quraysh family of the Banu Imayyah and the only member of it to have embraced Islam in the years of the vigil.[21] After the election of Othman, the noble and wealthy Banu Umayyah had one of their own at the summit of the Muslim community. Despite the Caliph's very weak personality, the twelve years of Othman's Caliphate witnessed the development and the continuation of the expansionist policy of Islamism, which was begun during Omar's decade.[22] After Othman's death, Ali ibn Abi Talib became Caliph. He was the cousin and son-in-law of the Prophet and ruled the Arabian Empire from 656 to 661 AD.[23] The Umayyad governor of Syria was Muawiyah ibn Abu Sufyan. There was some misunderstanding between Muawiyah and Ali.[24]

Muawiyah, the son of Abu Sufyan, inherited his father's intelligence, energy, and flair for politics. By his outstanding leadership he transformed Syria into a province to be used by the entire Muslim Empire as a model.[25] Muawiyah succeeded in founding a dynasty in his family, the Banu Umayya, who held the Caliphate and the Empire for ninety years.[26]

The Umayyad Caliphate

Muawiyah was proclaimed Caliph in 661 AD and ruled until 680 AD. With his accession the seat of the provincial government of Damascus became the capital of the Muslim Empire.[27] In his reign the consolidation as well as the extension of the territories of the Caliphate was established. He abolished many traditional features of the government and the earliest Byzantine framework, and he also developed a stable, well-organized state. Out of seeming chaos, he founded an orderly Muslim society.[28]

In Muawiyah the sense of *finesse politique* was developed to a degree probably higher than in any other Caliph. To his Arab biographers, his supreme virtue was his *hilm*, that unusual ability to resort to coercion only when it was absolutely necessary and to use peaceful measures in every other instance.[29] As Hitti stated:

> His prudent mildness by which he tried to disarm the enemy and shame the opposition, his slowness to anger, and his absolute self-control left him under all circumstances master of the situation. 'I apply not my sword where my lash suffices, nor my lash where my tongue is enough. Even if there be one hair binding me to me fellowmen, I do not let it break: when they pull I loosen, and if they loosen, I pull.'[30]

The next important Caliph Marwan (683-5 AD), the founder of the Marwanid branch of the Umayyad dynasty, was his son 'Abd-Al-Malik (685-705 AD), the 'father of Kings.' Under Abd-Al-Malik's rule and that of his four sons who succeeded him, the dynasty at Damascus reached the zenith of its power and glory. During the reign of Al-Walid and Hisham, the Islamic Empire reached its greatest expansion, stretching from the shores of the Atlantic Ocean and the Pyrenees to the Indies and the coast of China. This was hardly rivalled in ancient times and was surpassed in modern times only by the British and Russian empires.[31]

The reign of Al-Walid (705-15 AD) was begun by an official changing of language of the public registers (*diwan*) from a multitude of tongues to Arabic. From this, the translation of difference scientific material started.[32] In 747 AD, open revolt against the Umayyads was proclaimed by their cousins, the

Abbasids, descendants of an uncle of the Prophet, Al-Abbas. After the success of the Abbasids, the Umayyad house was exterminated.[33]

The Abbasid Caliphate

The Umayyad empire was Arabian in nature; the 'Abbasid was more international. The 'Abbasid was an empire of Neo-Muslims in which the Arabs formed only one of the many component races.[34] Like other dynasties in Muslim history, the 'Abbasid dynasty attained its most brilliant period of political and intellectual life soon after its establishment. The Baghdad Caliphate, founded by Al-Saffah and Al-Mansur, reached its height in the period between the reigns of the third Caliph, Al-Mahdi, and the ninth, Al-Wathiq, and more particularly in the days of Harun Al-Rashid and his son, Al-Ma'mun.[35]

It was mainly because of the illustrious and brilliant Caliphs, Harun Al-Rashid and Al-Ma'mun, that the 'Abbasid dynasty acquired a 'halo in popular imagination' and became the most celebrated in the history of Islam.[36] Following the rule of Al-Wathiq, the state began to decline until the Caliph Al-Musta'sim, the thirty-seventh in succession, met final destruction at the hand of the Mongols in 1258. An idea of the degree of power, glory, and progress attained by the 'Abbasid Caliphate at its highest and most promising period may be gained from observing the security of its foreign relations, the court and aristocratic life in its capital, Baghdad, and the unparalleled intellectual awakening which culminated under the patronage of Al-Ma'mun.[37] Under the rule of Harun Al-Rashid began the translation of classical mathematics in Greek and Sanskrit to Arabic and, in general, an overall increase in mathematical activity. His son, Al-Ma'mun, who succeeded him, was a patron of mathematics and an astronomer.[38] Under his influence, systematic mathematical study was begun in the ninth century.

Among the mathematicians of this period was Al-Khwarizmi, the man who influenced mathematical thought to a greater extent than any other medieval writer. In addition to compiling the oldest astronomical tables, Al-Kharizmi composed the oldest work on arithmetic and the oldest work on algebra. These were translated into Latin and used until the sixteenth century as the principal mathematical textbooks by European universities. They also served to introduce into Europe both the science of algebra and its name. Al-Khwarizmi's work was also responsible for introducing Arabic numerals to the Western culture.[39]

The 'Abbasid period is remembered as a brilliant and prosperous era in the history of Islam. The rulers distinguished themselves as great patrons of learning and under their influence scholars contributed considerably to the advancement of world civilization.[40] Under this liberal dynasty in Baghdad, the great movement of Arabic science flourished and opened the way for 'Islam's Golden Age.'[41]

The Muslims in Europe

While the Eastern branch of the Muslim Empire was approaching its golden day, the far-Western branch was enjoying a period of corresponding splendor. The Muslim West, absorbing elements of Christian culture of the early Middle Ages, produced the civilization which was later to be remembered as the Golden Age of Muslim culture.[42]

The Muslims in Spain

It was in 750 AD that the Umayyad dynasty in Damascus was overthrown by the 'Abbasid family. Among the few who escaped was a youth of twenty, Abd-Al-Rahman, who possessed courage and leadership ability. Making his way to Spain, he fought to reach power and maintain the Umayyad dynasty which was destroyed in the East.[43] He initiated the intellectual movement which made Cordova a center of world culture.[44] The Umayyad dynasty remained until the tenth century when civil disturbances, tribal revolts, and general political incompetence on the part of the Amirs reduced the organized Muslim state of Spain to the city of Cordova and its environs.[45]

The reign of Abd-Al-Rahman and that of his two immediate successors marked the height of Moslem rule in the West, and during this period or about the tenth century, the Umayyad capital of Cordova took its place as the most cultured city in Europe.[46]

Al-Hakam, successor to Abd-Al-Rahman III, was himself a scholar and patron in learning. He granted lavish subsidies to scholars and established twenty-seven free schools in the capital. Under him, the University of Cordova, founded in the principal mosque by Abd-Al-Rahman III, became a place of pre-eminence among the educational institutions of the world. It preceded Karawiyin University in Fez in Morocco, Al-Azhar of Cairo, and the Nizamiyah of Baghdad, and attracted students, both Christian and Moslem, from Spain and other parts of Europe, Africa, and Asia.[47]

The Muslims in Sicily

The only area in Europe other than Spain in which the Muslims gained a firm foothold was Sicily.[48] The Muslim conquest of Sicily, which had begun with periodic raids as early as 652 AD, had been completed in 827. During the next 189 years, under the rule of Muslim chieftains, Sicily was transformed into a province of the Muslim world with Palermo as its capital.[49]

By being at the meeting point of two cultural areas, Sicily became a medium for transmitting ancient and medieval culture.[50] Its population was comprised of a Greek element which used Greek, an Arabian element which spoke Arabic, and a group of scholars who knew Latin. All three languages were in concurrent use in the official registers and royal charters as well as among the populace of Palermo. The main contribution to the Muslim culture was the translation of Greek writings which dealt mainly with astronomy and mathematics. Although some of these Greek and Arabic works were translated in Toledo in Spain, Sicily's contribution was of prime value nonetheless.[51]

The Muslim Dilemma

After six centuries of the Umayyad and the 'Abbasid rule, the Muslim Empire experiençed a fity-year period of gradual political breakdown in which fragmentation eventually overcame unity. This political deterioration set the stage for the invasion of the Empire in 1258 AD by the Mongols under Hulagu Khan, the grandson of Genghis Khan, who had ravaged Asia and terrorized Europe.[52] The Mongols were ruthless warriors and their guiding creed was expressed by Genghis Khan in these words:

> The greatest joy is to conquer one's enemies, to pursue them,

to seize their property, to see their families in tears, to ride their horses, to possess their daughters and wives.[53]

When Hulagu Khan and his men swept across Baghdad, the powerless 'Abbasid Caliph surrendered after a weak defence. As an indication of his scorn, Hulagu had him put in a sack and trampled to death. Although suffering less than some other cities, Baghdad was plundered, priceless libraries and works of art were destroyed, and many of the inhabitants were massacred. The Mongols continued their destruction elsewhere in Mesopotamia and Syria, and the great system of irrigation which had made the region fertile and prosperous for thousands of years, was ruined.[54]

The Ottoman Turks

The Ottoman Turks first appeared in Asia Minor in the thirteenth century as a frontier tribe on the western confines of the Seljuk Sultanate of Rum (or Iconium). Better disciplined and organized than their immediate neighbours, they began to expand at the expense of the Seljuks and the Byzantines. The Seljuk Sultanate, weakened by Mongol pressure, soon disintegrated into petty principalities which fell under the Ottomans.[55] In the fourteenth century the Ottomans established themselves at strategic points in Greece, Serbia, and Bulgaria. By the middle of the following century the Byzantine Empire was almost surrounded by Ottoman possessions and vulnerable to a major attack which came with the conquest of Constantinople by the Turks in 1453.[56]

By 1473 Asia Minor was firmly under Ottoman rule. Under Mohammed the Conqueror (1451-81), the Turks pushed their conquests further into Europe and Asia.[57] Under Mohammed's successors the Empire expanded to the East, while during the reign of Selim I (1512-20) the Turks conquered North Mesopotamia, Egypt, Syria, and part of the Arabian peninsula. The conquest of Egypt was important because it ended the 'Abassid Caliphs. The story that the last 'Abassid ruler transferred the title to the victorious sultan is not supported by historical evidence, but Ottoman rulers took the title. The Caliphate implied spiritual and temporal authority over the Muslims.[58]

The penetration of the Mediterranean world by the Turks was by no means a boon to eleventh-century Muslim civilization. The level of Turkish culture was much more primitive than that which prevailed among the sophisticated Arabic-speaking peoples of the eastern Mediterranean.[59]

Summary

Some Muslim historians claim that the geographic expansion of Arabian culture was an evolution rather than a predetermined action. Evidences in the *Qur'an* and in early political practices in Islam, both in the East and in the West, suggest that the religion of Islam aspired to expand in time into a major political system as well as a universal faith of mankind.[60] Mohammed sent emissaries to the King of Ethiopia and the emperors of Iran and Byzantium, inviting them to adopt Islam. In less than a century, Muslim influence extended its economic, political, and religious power over a territory matched in magnitude only by the Roman Empire. The faith of Islam spread in a short period of time over three continents, persuading diverse peoples and cultures from Spain to China either to adopt the new faith or accept its political jurisdiction. This explains the partial success of Arabia's original concept of life. Islam labored to bring the world under one system of religion, one form of government, and one way of life.[61] The Muslims' experiment in building an empire was more dramatic than enduring and, in reality, the Muslims expressed their genius in the building of a religion with widespread appeal rather than in the construction of a political order ruled by common consent.[62]

It shall be the purpose of the remaining chapters of this book to present an in-depth discussion of the Muslim development of the various branches of mathematics. The following chapter is devoted to an extensive study of arithmetic, one of the principal topics to be studied by the Muslims.

Notes

1. Andrew Crichton, *The History of Arabia: Ancient and Modern* (New York, Harper and Brothers, 1837), Vol. I, p.17.
2. Samuel Graham Wilson, *Modern Movements Among Moslems* (New York, Fleming H. Revell Company, 1916), pp. 105-9.

3. Sania Hamady, *Temperament and Character of the Arabs* (New York, Twayne Publishers, 1960), p. 19.
4. Rev. George Bush, *Life of Mohammed: Founder of the Religion of Islam and the Empire* (Niagara, Henry Chapman, 1831), p. 14.
5. Francesco Gabrieli, *The Arabs: A Compact History* (New York, Hawthorn Books, 1963), p. 26.
6. Al-Imam Abu 'Abdullah Muhammad B. Isma'il Al-Bukhari, *At-Ta-Rikhu'l-Kabir* (A Dictionary of the Biography of Traditionists) (Hyderabad, India, Osmania Oriental Publications Bureau, 1942), Vol I, Part I, pp. 5-10.
7. Gustave E. Von Grunebaum, *Medieval Islam: A Study in Cultural Orientation*(The University of Chicago Press, 1947), p. 71.
8. Nabih 'Agil, *Tarikh Al-Arab Al-Gadim* (Damascus University Press, 1968), p. 551.
9. W. T. Sedgwick and H. W. Tyker, *Short History of Science* (New York, The Macmillan Company, 1925), p. 160.
10. Ralph Linton, *The Tree of Culture* (New York, Afred Knopf, 1955), p. 378.
11. George Sarton, *The Life of Science: Essays in the History of Civilization* (New York, Henry Schumann, 1948), p. 146.
12. A. S. Tritton, *Islam: Beliefs and Practices* (London, Hutchinson's University Library, 1951), p. 109.
13. Rom Landau, *Islam and the Arabs* (New York, The Macmillan Company, 1959), pp. 40-1.
14. 'Abi Al-Hasan Ali ibn Muhammed Al-Ma'ruf bi'ibn Al-'Athir, *Al-Kamil fi Attarikh* (Cairo, 'Idarat Attiba'at Al-Muniriyah bi Masr, 1929), Vol. II, p. 222.
15. Sir John Bagot Glubb, *The Great Arab Conquests* (Englewood Cliffs, New Jersey, Prentice-Hall, Inc, 1964), p. 106.
16. Kenneth W. Morgan, *Islam – The Straight Path* (New York, The Ronald Press Company, 1958), p. 49.
17. John Joseph Saunders, *A History of Medieval Islam* (London, Routledge and Kegan Paul, 1966), p. 44.
18. Joel Carmichael, *The Shaping of the Arabs: A Study in Ethnic Identity* (London, Collier Macmillan, 1967), pp. 88-9.
19. William Stearns Davis, *A Short History of the Near East: From the Founding of Constantinople* (New York, The Macmillan Company, 1922), p. 132.
20. Arnold J. Toynbee, *A Study of History* (London, Oxford University Press, 1939), Vol. III, p. 466.
21. Carl Brockelmann, *History of the Islamic Peoples* (Cornwall, New York, The Cornwall Press, 1947), p. 63.
22. Gustave Edmund von Grunebaum, *Modern Islam: The Search for Cultural Identity* (Berkeley and Los Angeles, University of California Press, 1962), pp. 119-20.
23. Ahmid Zaki Sufwat, *Jamharat Rusa'il Al-'Arab fi 'usur Al-'Arabiyah Azzahirah* (Cairo, Sharikat Maktabat wa Mutba'at Mustafa Al-Babi Al-Halabi wa 'Wladuh bi Masr, 1937), Vol. I, pp. 479-82.
24. Fazlur Rahman, *Islam* (New York, Holt, Rinehart, and Winston, 1966), p. 171.
25. Arnold Hottinger, *The Arabs: Their History, Culture, and Place in the*

Modern World (Berkeley and Los Angeles, University of California Press, 1963), p. 46.
26. John Bagot Glubb, *The Life and Times of Mohammed* (New York, Stein and Day Publishers, 1970), p. 368.
27. John Bagot Glubb, *The Course of Empire: The Arabs and Their Successors* (London, St. Paul's House, 1965), p. 26.
28. Bernard Lewis, *The Arabs in History* (New York, Hutchinson's University Library, 1950), p. 65.
29. Amir Ali Sayed, *Mukhtasar Tarikh Al-'Arab* (Beirut, Dar Al-Galam Llmalayin, 1961), p. 88.
30. Philip K. Hitti, *Makers of Arab History* (New York, Harper and Row Publishers, 1971), p. 43.
31. Hamilton Alexander Rosskeen Gibb, *Studies of the Civilization of Islam* (Boston, Beacon Books on World Affairs, 1962), p. 35.
32. Philip K. Hitti, *The Arabs: A Short History* (London, Macmillan, 1948), p. 217.
33. Sir Hamilton Alexander Rosskeen Gibb, *An Interpretation of Islamic History* (Lahore, M. Ashraf Darr for Oriental Publishers, 1957), p. 12.
34. Gustave Edmund von Grunebaum, *Islam: Essays in the Nature and Growth of a Cultural Tradition* (London, Routledge and Kegan Paul, 1961), p. 16.
35. Rangrut, *The Ideologies in Conflict* (Karachi, Central Printing Press, 1964), p. 107.
36. Joseph Hell, *The Arab Civilization* (London, W. Heffer and Sons, 1943), pp. 72-3.
37. Alfred Guillaume, *Islam* (Edinburgh, R. and R. Clark, 1954), pp. 82-3.
38. Edna E. Kramer, *The Main Stream of Mathematics* (New York, Oxford University Press, 1951), p. 22.
39. 'Uthman Al-Ka'ak, *Al-Haddarah Al-'Arabiyah fi Hawdd Al-Bahr Al-'Abyad* (Cairo, Jami'at Adduwal Al-'Arabiyah, 1965), pp. 73-114.
40. Hassan Ibrahim Hassan, *Islam: A Religious, Political, Social, and Economic Study* (Baghdad, Iraq, The Times Printing and Publishing Company, 1967), p. 132.
41. Lewis Samuel Feuer, *The Scientific Intellectual: The Psychological and Sociological Origins of Modern Science* (New York, Basic Books, 1963), p. 184.
42. Muhammed Mahdi Al-Basiri, *Al-Mawashah fi Al'Andalus wa fi-Asharg* (Baghdad, Mutba'at Al-Ma'arif, 1948), pp. 4-5.
43. Khalid Assufi, *Tarikh Al-'Arab fi 'Asbanya* (Damascus, Al-Mutba'ah Atta'awiniyah, 1959), pp. 3-7.
44. Lane Poole, *The Story of the Moors in Spain* (London, G. P. Putnam's Sons, 1902), pp. 33-45.
45. Budgett Meaking, *Moorish Empire: A Historical Epitome* (London, Swan Sonnenschein and Co, 1899), pp. 45-6.
46. R. H. Towner, *The Philosophy of Civilization* (New York, G. P. Putnam and Sons, 1923), p. 117.
47. Khalid Assufi, *Tarikh Al-'Arab fi 'Sbanya: Nihayah Al-Khilafah Al-'Amawiyah fi Al-'Ndalus* (Halabb, Maktabat Dar Ashsharg, 1963), pp. 334-5.
48. Historical Section of the Foreign Office, *Mohammedan History: The Rise of Islam and the PanIslamic Movement* (London, H. M. Stationery

Office, 1920), Vol. X, p. 22.
49. Samuel M. Zwemar, *Islam* (New York, Student Volunteer Movement for Foreign Missions, 1907), pp. 65-6.
50. Arnold J. Toynbee, *Civilization on Trial* (New York, Oxford University Press, 1948), p. 185.
51. Philip K. Hitti, *History of the Arabs: From the Earliest Time to the Present* (London, Macmillan, 1964), pp. 612-3.
52. Charles F. Gallagher, *A Note on the Arab World* (New York, American University Field Staff, 1961), p. 16.
53. Arabian American Oil Company, *Aramco Handbook: Oil and the Middle East* (Netherlands, Joh. Enschede en Zonen-Haarlem, 1968), p. 40.
54. Ibid.
55. F. R. J. Verhoeven, *Islam* (New York, St. Martin's Press, 1962), p. 51.
56. Hamilton Alexander Rosskeen Gibb and Harold Bower, *Islamic Society and the West* (London, Oxford University Press, 1950), Vol. I, Part I, pp. 42-3.
57. Hamilton Alexander Rosskeen Gibb and Harold Bowen, *Islamic Society and the West* (London, Oxford University Press, 1957), Vol. I, Part II, p. 70.
58. George Lenczowski, *The Middle East in World Affairs* (Ithaca, New York, Cornell University Press, 1956), p. 4.
59. Norman F. Canot, *Medieval History: The Life and Death of a Civilization* (New York, The Macmillan Company, 1963), p. 274.
60. Sa'id Nasir Addahan, *Al-Jur'an wa Al'Ulum* (Karbala, Mutba'at Anna'man, 1965), p. 169.
61. Lajid Khadduri, *The Law of War and Peace in Islam* (London, Luzac and Company, 1941), p. 9.
62. Gustantin Zurig, *Fi Ma'rakati Al-Hadarah* (Beirut, Dar Al-Galam Li Lmalayin, 1963), pp. 77-8.

3 ARITHMETIC

Mathematics, the man-made universe, appears to have emerged from man's primitive needs to keep records, to communicate information, and to understand and control his environment. Certainly, arithmetic was among the first branches of mathematics to develop and flourish as the concepts of number and operations on numbers came into general usage. This development was no doubt a gradual one, but the advantages of counting soon led to the improvement and extension of basic mathematical concepts, which have proliferated over the centuries into what we today refer to as number theory.

It is believed that arithmetic came into existence before written language developed. Thus, the history of mathematics, stemming from arithmetic, is a part of the history of civilization. Moreover, the progress of man over the era of recorded history is largely paced by his use and grasp of mathematical ideas. The use and manipulation of symbols as mental representations of physical things led to the idea of the early use of mathematical operations of addition and subtraction without the need to count the real objects in a set.[1]

> Arithmetic is the foundation of all mathematics, pure or applied. It is the most useful of all sciences, and there is, probably, no other branch of human knowledge which is more widely spread among the masses.[2]

Among the Muslim mathematicians who contributed the most to arithmetic was Abu-Yusef Ya'qub ibn Ishaq Al-Kindi.[3] Al-Kindi was born about 801 AD at Kufah during the governorship of his father. This position was occupied earlier by his grandfather. The surname indicates ancestry in the royal tribe of Kindah of Yamanite origin. Al-Kindi is known in the West as Alkindus.[4] To his people he became known as Faylusaf Al-Arab,[5] the philosopher of the Arabs. He was the only notable philosopher of pure Arabian blood and the first one in Islam. Al-Kindi 'was the most learned of his age, unique among his contemporaries in the knowledge of the totality of ancient sciences, embracing logic, philosophy, geometry, mathematics, music, and astrology.'[6] According to Professor Emeritus Philip K. Hitti of Princeton University:

> He [Al-Kindi] was a man with a firstclass mind which addressed itself to the study of the new philosophy. It was an encyclopaedic mind to which no aspect of human knowledge seemed alien.[7]

Among his contributions to arithmetic, Al-Kindi wrote eleven texts on the subject. The following is a composite listing of the titles of these works:

1. An Introduction to Arithmetic
2. Manuscript on the Use of Indian Numbers
3. Manuscript on Explanation of the Numbers mentioned by Plato in his politics
4. Manuscript on the Harmony of Numbers
5. Manuscript of Unity from the Point of View of Numbers
6. Manuscript on Elucidating the Implied Numbers
7. Manuscript on Prediction from the Point of View of Numbers
8. Manuscript on Lines and Multiplication with Numbers
9. Manuscript on Relative Quantity
10. Manuscript on Measuring of Proportions and Times
11. Manuscript on Numerical Procedures and Cancellation[8]

Al-Karkhi of Baghdad (1020 AD) was the most scholarly and the most original writer of arithmetic. Two of his works are known. The first is the *Al-Kafi fi al-Hisab* (*Essentials of Arithmetic*), which gives the rules of computations. His second work, *Al-Fakhri*, derives its name from Al-Karkhi's friend, the grand vizier in Baghdad at that time.[9]

A Latin translation of a Muslim arithmetic text was discovered in 1857 at the Library of the University of Cambridge. Entitled *Algoritmi de numero Indorum*, the work opens with the words: 'Spoken has Algoritmi. Let us give deserved praise to God, our leader and defender.'[10] It is believed that this is a copy of Al-Khwarizmi's arithmetic text which was translated into Latin in the twelfth century by an English scholar. Before it was lost, this translated version of Al-Khwarizmi's text found its way to Italy, Spain, and England. Its name, through various modifications, became Alchwarizmi, Al-Karismi, Algoritmi, Algorismi, which named the new art, Algorithm.[11] Thus, Al-Khwarizmi left his name to the history of mathematics in the form of Algorism, the old word for arithmetic.[12]

Arabic Numerals

Imagine a hillside thousands of years ago. A man emerges from a cave. His brow is heavy, and his arms are long and muscular. Around his waist he wears a tattered animal skin, and a herd of wild horses passes below him. Back into the cave he rushes and, with grunts and gestures, excitedly tells his clan that 'many, many' horses are passing. This is the best he can count. He has no way of telling them that 30, 40 or 50 horses are in the herd, for at best he knows three numbers – one, two, and 'many'. Civilizations will rise and fall, and even his own form will change before man learns to count with the ease and exactness of numbers such as 30, 40 or 50. The development of an easy-to-use, easy-to-learn system of numbers was a milestone, reached only after long struggle. In fact, man has had such a system only for about 1,000 years, and mankind has been on earth for a very long time.

In every civilization of which there is historical record, there exists some idea of numbers.[13] In the early and more primitive civilizations, this concept is exhibited in a set of number symbols or words.[14] It is common knowledge that the numerals in which the score is given at a football game are called Arabic numerals and one assumes that these have always been in use. In actual fact Europe adopted them from the Muslims only in the thirteenth century. Fighting their introduction and that of the decimal system that went with them for several hundred years, Europe deprived itself of the advantages of one of the world's greatest contributions to mathematics.

Prior to the Arabian numerals, the West relied upon the clumsy system of Roman numerals, and before that upon the even more clumsy Greek numerals. In the decimal system, the number 1843 can be written in four numerals, whereas in the Roman numerals, eleven figures are needed. The result is MDCCCXLIII. It is obvious that even for the result of the simplest arithmetical problem, Roman numerals called for an enormous expenditure of time and labor. The Arabian numerals, on the other hand, render even complicated mathematical tasks relatively simple.[15] Professor J. Houston Banks of Peabody College has stated:

> The Roman system seems to have some advantage over the present numerals when we consider the process of addition. Let us consider the addition of 127 and 58 in Roman numerals:
>
> C X X V I I
> L V I I I
> C L X X V V I I I I I
>
> We can merely bring down all the symbols in each addend; then if we remember that five I's are written as V and two V's as X, the answer becomes CLXXXV. It is not necessary to know the addition combination such as 7 + 8 and 5 + 2. However, the process is much longer and more cumbersome. But if we attempt to multiply or divide, it is a different story.[16]*

At the time of the Prophet Mohammed, Arabians had a script which did not differ significantly from that in later centuries. The letters of the early Arabic alphabet were then used as numerals among the Arabians. Arabic alphabetic numerals used before the introduction of the Hindu-Arabic numerals are listed in Figure 3.1.[17]

The numerals 1, 2, 3, 4, 5, 6, 7, 8, and 9 used in almost all parts of the globe are believed by some scholars to be related to nine Sanskrit characters used in ancient times by the Hindus. These numerals were transmitted to Muslims, who modified and introduced them into Europe.[18] The origin of these numerals, which the

---

* In reading this quotation one has, of course, to be aware that supposition value of Roman numerals is to be considered. The sum of XI and IX is neither XXII which would be 22 nor IIXX which a Roman student would have considered as a funny mispelling of XVIII.

Fig. 3.1: Early Arabian Numeration using Alphabetic Symbols

1	ا	10	ي	100	ق	1000	غ	10000	يغ	100000	قغ
2	ب	20	ك	200	ر	2000	بغ	20000	كغ	200000	رغ
3	ج	30	ل	300	ش	3000	جغ	30000	لغ	300000	شغ
4	د	40	م	400	ت	4000	دغ	40000	مغ	400000	تغ
5	ه	50	ن	500	ث	5000	هغ	50000	نغ	500000	ثغ
6	و	60	س	600	خ	6000	وغ	60000	سغ	600000	خغ
7	ز	70	ع	700	ذ	7000	زغ	70000	عغ	700000	ذغ
8	ح	80	ف	800	ض	8000	حغ	80000	فغ	800000	ضغ
9	ط	90	ص	900	ظ	9000	طغ	90000	صغ	900000	ظغ

Muslims themselves call 'Hindi' numerals, is somewhat uncertain and vague. Some writers have suggested that the word 'Hindi' does not necessarily imply that numbers originated in India, since the Arabians had many meanings for the word. It is perhaps significant that the first Arabic book containing 'Arabic' numerals was written in 874 AD, while the first Indian book containing them appeared two years later.[19]

The Muslim culture brought to the world two great streams of human achievement, a new number language from the East, and the classical mathematics of the West.[20] Nonetheless, whether the numbers were discovered by the Indians or the Muslims, it is indisputably the Muslim mathematicians who made use of them and introduced the decimal system to the world. The ingenious idea of expressing all numbers by means of ten symbols, with each symbol receiving the value of position as well as an absolute value, had escaped both the Alexandrians and the Greeks.

The Arabian system of numeration, based upon the idea of place value, is one of the most rewarding results of human intelligence and deserves the highest admiration. This simplicity of numeration is one of the greatest achievements of the human mind. In the hands of a skilled analyst, it becomes a powerful instrument in 'wresting from nature her hidden truths and occult laws.'[21] Lee Emerson Boyer states:

> Without it, many of the arts would never have been dreamed of, and mathematics would have been still in its cradle. With it, man becomes armed with prophetic power, – predicting eclipses, pointing out new planets which the eye of the telescope had not seen, assigning orbits to the erratic wanderers of space, and even estimating the ages that have passed since the universe thrilled with the sublime utterance, 'Let there be light!' Familiarity with it from childhood detracts from our appreciation of its philosophical beauty and its great practical importance. Deprived of it for a short time, and compelled to work with the inconvenient method of other systems, we should be able to form a truer idea of the advantages which this invention has conferred on mankind.[22]

A very powerful system of numeration is used for arithmetic

computations. The system evolved slowly and received contributions from many civilizations. It is known, however, as the Arabic system of numeration because the Muslims made major contributions. A study of the Arabic system of numeration is worthwhile for the reasons that one will be able to:

1 witness and appreciate the 'beauty' and logic of the system;
2 gain a better understanding and knowledge of the system used since childhood; and
3 experience the effect that the 'base' has upon this system.[23]

Muslims Offered the Zero

There is no numeral in the set of Arabic numerals of greater significance than the zero, which is referred to as *sifr* or 'empty' by the Arabians. While the zero is used as a symbol for nothing, it actually has much meaning. The difference in appearance between 5 and 50 is only a zero, but that small circle is actually one of the world's greatest mathematical innovations. In combination with the nine basic numerals, the zero provides numbers with an infinite variety of values. The zero's creation opened the way for the entire concept of algebraic positive and negative numbers, which are used for calculations, identification of electrical charge and discharge, for navigation, etc.

It is interesting to note that while the earliest Hindu example of a zero was found on an inscription of 876 AD at Gwalior, the earliest Muslim zero is contained in a manuscript dated 873 AD. Without the *sifr*, any number system would be much more complicated and clumsy. It took Europe at least two hundred and fifty years to accept and acknowledge the zero as a gift from the Muslims. The concept of a mathematical representation that appeared to have no content of its own made no sense to European mathematicians. It was not until the late twelfth century that the Europeans really began to make use of the zero and the decimal system.[24]

The Hindus considered a position 'empty' if it were not filled out, and, thus, the word *sunya* (empty or blank) was used for zero.[25] The Muslims translated the Indian *sunya* as *sifr*. When Fibonacci (Leonardo of Pisa) wrote his *Liber Abaci* in 1202, he spoke of the symbol as *Zephirum*. A century later Maximus Planudes (1340) called it *Tzipha,* and this form was still used as late as the sixteenth century. In Italian it was called *Zenero, Cenero,* and *Zephiro.* Since the fourteenth century, zero has been the term used as shown in the records of 1491 by Calnadri and of 1494 by Luca Pacisli. The word *nulla* appears in Italian translations of Muslim writings of the twelfth century, and also in the French (Chuquet, Triparty) and German of the fifteenth century. *Cipher* was still used for zero by Adrian Metiers (1611), Herigone (1634), Cavalieri (1643), and Euler (1783), even though the more modern German word *Ziffer* had been introduced. The zero symbol is also called *cipher* and *naught,* and modern usage permits it to be called 0 (word '0') 'an interesting return to the Greek name *omicron.*'[26]

Until the intervention of the symbol for zero, it was necessary to have paper or tablets in columns, in order to keep the digits in their proper places. Thus, the numbers shown in Figure 3.2 would stand respectively for 203, 4020, and 100. The purpose of a zero is to keep the other digits in their proper position.[27]

Figure 3.2: Positional Method

	2		3
4		2	
	1		

The Hindu symbol for the zero was '⊙' (a dot inside a circle). However, in the Muslim Empire, the Muslim East (Asia, Baghdad) and the Muslim West (North Africa, Spain), a number of different representations were used. In the Muslim East, the dot '•' was used and their augmented sets of numerals became: ١, ٢, ٣, ٤, ٥, ٦, ٧, ٨, ٩, ٠. The Muslim West adopted the circle ○ as their symbol and the complete set of numerals became 1, 2, 3, 4, 5, 6, 7, 8, 9, and 0.[28]

Through the efforts of Muslim scholars, the decimal numerational system was introduced to the civilized world. It was a system in which the zero was indeed the pivotal point and the elementary operations of arithmetic were tremendously simplified.[29]

Operation

The Muslims adopted the Greek definition of arithmetical operations but used techniques of their own. Some of the procedures may seem complicated and somewhat unusual because they were based on prior analysis of number construction.[30]

Multiplication

The Hindus' multiplication was very complicated. For instance, for the problem 569 times 5, they generally said:

> 5 . 5 = 25; 5 . 6 = 30, which changed 25 into 28;
> 5 . 9 = 45. Hence the 0 must be increased by 4. The product is 2845.[31]

However, the Muslims' multiplication was very simple and more readily performed. They used the network or lattice method (*shabacah*) in which the tablet was divided into squares resembling a chess board. Diagonals were drawn.[32] The multiplication of 239 × 567, is illustrated in Figure 3.3. To find the product by the use of this device one proceeds as follows. The factors are placed at the top and left of the rectangle. The product for each cell is formed by taking the product of the row and column element for it and recording the digit above and the tens digit below the diagonal. The product of the two original factors is determined by summing the numbers in each diagonal and 'carrying' if necessary.

Division

Fibonacci studied in Muslim schools and in 1202 AD introduced Arabian numerals to Europe.[33] He treated several cases in division, the first being division by a one-digit number. He divided 10,004 by 8 as an example by writing the quotient below the divisor and the remainder above the dividend:

Fig. 3.3: Multiplication by the Lattice Method

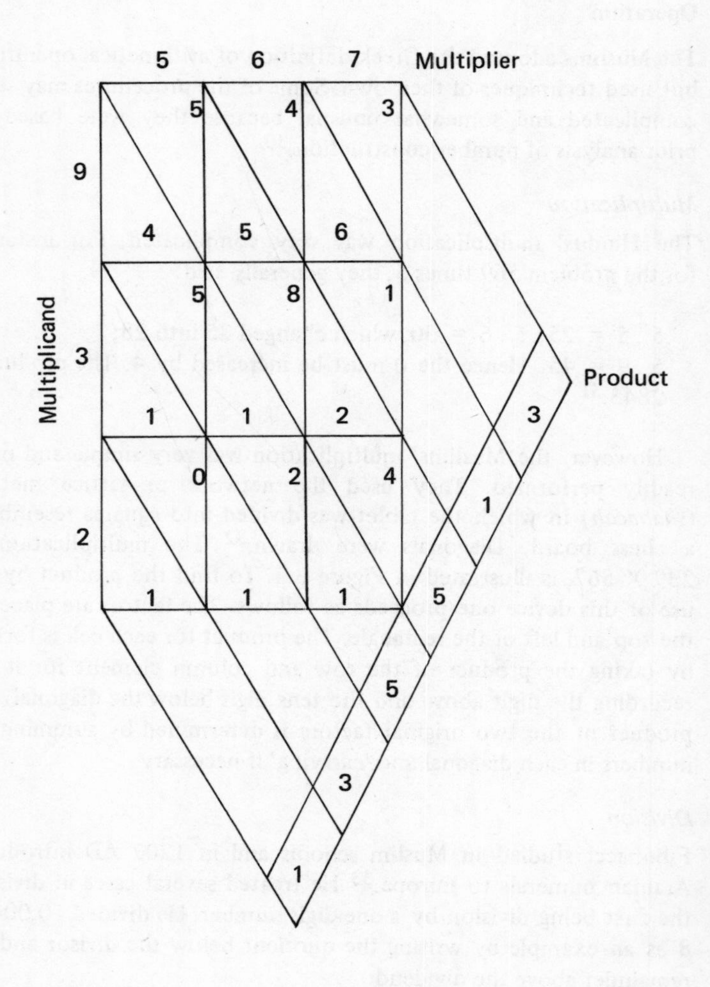

Fig. 3.4: An Example of the Arabian Long-Division Method

1	7	9	7	8
4	7	2		
		0		

A

1 1	7 2	9	7	8
	5 2	9 1	7	8
	3	8	7 6	8
4	3 4 7	8 7 2	1 2	8
		0	3	

B

1 1	7 2	9	7	8
	5 2	9 1	7	8
	3	8	7 6	8
	3 3	8 2	1	8
		6 5	1 6	8
			5 1	8 6
			4	2
			4	2
4	4 7	4 7 2	7 2	2
		0	3	8

C

4
10004
8
1250

Fibonacci advised one to divide the factors of a number whenever possible, and whenever the divisor is greater than 10, he suggested using the nearest multiple of 10 as a trial divisor. This was taken by him from the Muslims.[34]

The Muslim method of long division, which required the skill of a mathematical expert, is the oldest method of long division known in the Muslim Empire. An example is presented to illustrate the method. To divide 17978 by 472, a sheet of paper is divided into as many vertical columns as there are digits in the number to be divided. The number to be divided is written at the top of the page and the divisor at the bottom, the first digit of each number being placed at the left-hand side of the paper. Then, taking the left-hand column, 4 divides one zero times; hence, the first digit in the dividend is 0, which is written under the last digit of the divisor. This is represented in Figure 3.4-A. In Figure 3.4-B, the divisor 472 is rewritten immediately above its former position, but shifted one place to the right and the numerals are cancelled. Then 4 evenly divides 17 four times; but, as on trial, it is found that 4 is too large for the first digit of the dividend and 3 is selected. Therefore, 3 is written below the last digit of the divisor and next to the digit of the dividend last found. The process of multiplying the divisor by 3 and subtracting from the number to be divided is indicated in Figure 3.4-B and shows that the remainder is 3818. A similar process is then repeated, that is, 472 is divided into 3818, showing that the quotient is 38 and the remainder, 42. This is represented in Figure 3.4-C, which shows the completed process.

Fractions (Al-Kasr)

The earliest treatment of the subject of common fractions is the *Lilavati* (1150 AD) of the Hindu mathematician Bhaskara II. Common fractions in *Lilavati* are denoted by writing the numerator above and the denominator below without any line between them. For example, 3/11 was written as:

3
11

Mixed numbers were written with the integer part above the fraction: thus, 8¾ was written:

8
3
4

The introduction of the line of separation is due to the Muslims. To denote a fraction in the Muslim method, one writes three-fourths as ¾; and to denote 3 + ¾, one writes the mixed number '3¾'. It is to Muslim mathematicians that credit is due for the first use of decimal fractions. Louis Charles Karpinski wrote:

> Our notation of fractions is quite certainly based upon Arabic forms . . . The Arabic word for fraction, al-kasr, is derived from the stem of the verb meaning 'to break.' The early writers on algorism commonly used *fractio*, while Leonardo of Pisa and John of Meurs (fourteenth century) used both fractio and minutum ruptus.[35]

Amicable Numbers

A pair of numbers is said to be amicable when the sum of the factors of one is equal to the other, and conversely. Therefore, M and N are amicable numbers when $\sigma_0(N) = M$, and also $\sigma_0(M) = N$.[36] For example, the numbers 220 and 284 are a pair of amicable numbers. The factors of 284 are 1, 2, 4, 71, 142, and their sum is 220. The factors of 220 are 1, 2, 4, 5, 10, 11, 20, 22, 44, 55, 110, and their sum is 284.

In Muslim mathematical writings the amicable numbers occur repeatedly. They play a role in magic and astrology, in the casting of horoscopes, in sorcery, in talismans.[37] Amicable numbers were one of the hobbies of Abu Zaid Abdel Rahman ibn Khaldun, who was born in Tunis in 1332.[38] Ibn Khaldun wrote that persons who have concerned themselves with talismans affirm that the amicable numbers 220 and 284 have an influence to establish a union or close friendship between two individuals.[39]

The ninth century was a glorious one in Muslim mathematics, for it produced not only Al-Khwarizmi in the first half of the century, but also Thabit ibn Qurra (826-901) in the second half. If Al-Khwarizmi resembled Euclid as an 'elementator', then Thabit is the Arabic equivalent of Pappus, the commentator on higher mathematics.[40]

The famous Thabit ibn Qurra also worked on amicable numbers. He was the first writer who reached fame and recognition in this period, and is particularly noted for his translations from Greek to Arabic of works by Euclid, Archimedes, Apollonius, Ptolemy, and Eutocious.[41] Had it not been for his efforts, the number of Greek mathematical works known today would be smaller, as for example, there would have been preserved only the first four rather than the first seven books of Apollonius' *Conics.* Thabit had so thoroughly mastered the content of the classics he translated that he suggested modifications and generalizations, and a remarkable formula for amicable numbers is credited to him.[42] The formula is as follows:

If p, q, and r are prime numbers, and if they are of the form,

$$p = 3 \cdot 2^n - 1, \quad q = 3 \cdot 2^{n-1} - 1, \quad r = 9 \cdot 2^{2n-1} - 1,$$

then p, q, and r are distinct primes and $2^n pq$ and $2^n r$ are a pair of amicable numbers.[43]

For instance, for $n = 2$,

since $p = 3 \cdot 2^n - 1$
therefore $p = 3 \cdot 2^2 - 1 = 3 \cdot 4 - 1 = 11$
since $q = 3 \cdot 2^{n-1} - 1$
$= 3 \cdot 2^{2-1} - 1 = 5$
since $r = 9 \cdot 2^{2n-1} - 1$
$= 9 \cdot 2^{4-1} - 1 = 9\,(8) - 1 = 71$
and since the pair of amicable numbers are $2^n pq$, $2^n r$, it follows that, $2^2\,(11)\,(5) = 220$, and $2^2\,(71) = 284$.

Therefore, it is shown that 220 and 284 are amicable numbers.

Sum of Natural Numbers

Muslim mathematician, Al-Karkhi gave expressions for the sum of the first, second, and third powers of the first *n* natural numbers as follows:[44]

$$1 + 2 + 3 + \ldots + n = \frac{n(n+1)}{2}$$

$$1^2 + 2^2 + 3^2 + \ldots + n^2 = \frac{n(n+1)(2n+1)}{6}$$

$$1^3 + 2^3 + 3^3 + \ldots + n^3 = \frac{n^2(n+1)^2}{4}$$

Summary

The author belives that arithmetic is a vital part of daily living because it meets practical needs. R. L. Goodstein suggests that without numbers mathematical communication would be exceedingly cumbersome and complicated; but, a language lacking expressions for numbers can nonetheless express everything that can be said in the current medium of expression.[45] Conant feels that the Muslims disseminated their own numerical system in such a manner that their influence is detected without difficulty.[46]

Outstanding among the Muslim contributions to arithmetic was the development of and introduction into Europe of the Hindu-Arabic numeration system. The contribution of the place holder, zero, the introduction of the modern notation for common fractions, and the use of decimal fractions are only a few of the many contributions made to arithmetic by the Muslims.

Notes

1. John Desmond Bernal, *Science in History* (London, C. A. Watts and Company, 1957), p. 79.
2. Robias Dantzig, *Number, The Language of Science* (Garden City, New York, Doubleday and Company, 1956), p. 38.
3. David Eugene Smith and Louis Charles Karpinski, *The Hindu-Arabic*

Numerals (Boston, Ginn and Company, 1911), p. 10.
4. Ibn Al-Nadim, *Al-Fahrasat Li Ibn Al-Nadim* (Cairo, Al-Haj Mustafa Muhammed, 1800), pp. 371-2.
5. Yuhana Gamir, *Falasifat Al-Arab* (Beirut, Al-Muktabat Ash Shargiyah, 1957), p. 5.
6. Franklin Wesley Kokomoor, *Mathematics in Human Affairs* (New York, Prentice-Hall, 1945), p. 172.
7. Philip K. Hitti, *Makers of Arab History* (New York, Harper and Row Publishers, 1968), p. 187.
8. George N. Atiyah, *Al-Kindi: The Philosopher of the Arabs* (Karachi, Al-Karami Press, 1966), p. 185.
9. Oystein Ore, *Number Theory and its History* (New York, McGraw-Hill Book Company, 1948), p. 185.
10. Florence A. Yeldham, *The Story of Reckoning in the Middle Ages* (London, George C. Harrap and Company, 1926), p. 64.
11. Bodleian Library, Oxford, England, Marsh MSS, 489, fol. 145^r-166^r.
12. Charles Singer, *A Short History of Scientific Ideas to 1900* (London, Oxford University Press, 1968), p. 162.
13. David Eugene Smith, *Number Story of Long Ago* (Washington, D.C., The National Council of Teachers of Mathematics, 1962), p. v.
14. Howard Franklin Fehr, *A Study of the Number Concept of Secondary School Mathematics* (Ann Arbor, Michigan, Edwards Brothers, 1945), p. 14.
15. Jane Muir, *Of Men and Number: The Story of the Great Mathematicians* (New York, Dodd, Mead and Company, 1961), p. 28.
16. Houston Banks, *Elements of Mathematics* (Boston, Allyn and Bacon, 1969), 3rd edn., pp. 66-7.
17. Florian Cajori, *A History of Mathematical Notations* (La Salle, Illinois, The Open Court Publishing Company, 1928), Vol. I, p. 29.
18. Mayme I. Logsdon, *A Mathematician Explains* (Chicago, Illinois, The University Press, 1935), p. 43.
19. Abdel Salam Said, 'We Remember that Western Arithmetic and Algebra Owe Much to Arabic Mathematicians,' *Arab World*, V, Nos. 1-2, (January-February 1959), 5.
20. Lancelot Hogben, *Mathematics for the Millions* (New York, W. W. Horton and Comapny, 1946), p. 235.
21. Lee Emerson Boyer, *Mathematics: A Historical Development* (New York, Henry Holt and Company, 1949), pp. 29-31.
22. Ibid.
23. Donald Merrick, *Mathematics for Liberal Arts Students* (Boston, Princle, Weber and Schmidt, 1970), p. 104.
24. Rom Landau, *Arab Contribution to Civilization* (San Francisco, The American Academy of Asian Studies, 1958), p. 29.
25. C. B. Boyer, 'Zero: The Symbol, The Concept, The Number,' *National Mathematics Magazine*, XVIII (1944), pp. 323-30.
26. Marie Haden, 'A History of Our Numerals and Decimal System of Numeration,' (unpublished Master's Thesis, George Peabody College for Teachers, 1931), p. 25.
27. A. Hooper, *The River Mathematics* (New York, Henry Holt and Company, 1945), pp. 13-14.
28. Tawfiq Al-Tawil, *Al-'Arab wa Al-'ilm* (Cairo, Maktabat Al-Nahdah Al-Misriyah, 1968), p. 61.

29. H. E. Slaughter, 'The Evaluation of Numbers – An Historical Drama in Two Acts,' *The Mathematics Teacher*, XXI (October 1928), pp. 307-8.
30. Rene Taton, *History of Science: Ancient and Medieval Science – From the Beginnings to 1450* (New York, Basic Books, 1963), Vol. I, p. 406.
31. Florian Cajori, *A History of Mathematics* (New York, The Macmillan Company, 1924), p. 90.
32. Robert W. Marks (ed.), *The Growth of Mathematics From Counting to Calculus* (New York, Bantam Books, 1964), p. 100.
33. Philip K. Hitti, *The Near East in History – A 5000-Year Story* (New York, D. Van Nostrand Company, 1960), p. 253.
34. Vera Sanford, *A Short History of Mathematics* (New York, Houghton Mifflin Company, 1930), pp. 100-1.
35. Karpinski, op. cit.
36. Oystein Ore, *Number Theory and Its History* (New York, McGraw-Hill Book Company, 1948), pp. 98-9.
37. Ibid.
38. Muhsin Mahdi, *The Khaldun's Philosophy of History: A Study in the Philosophic Foundation of the Science of Culture* (Chicago, The University of Chicago Press, 1954), p. 27.
39. Leonard Eugene Dickson, *History of the Theory of Numbers: Divisibility and Primality* (Washington, D.C., Press of Gibson Brothers, 1919), Vol. I, p. 36.
40. Carl B. Boyer, *A History of Mathematics* (New York, John Wiley and Sons, 1968), p. 258.
41. George Sarton, *A History of Science* (Cambridge, Mass., Harvard University Press, 1952), Vol. I, p. 446.
42. Boyer, op. cit.
43. Howard Eves, *An Introduction to the History of Mathematics* (New York, Holt, Rinehart and Winston, 1969), p. 220.
44. W. W. Rouse Ball, *A Short Account of the History of Mathematics* (New York, Dover Publications, 1960), pp. 159-60.
45. R. L. Goodstein, 'The Arabic Numerals, Numbers, and the Definition of Counting,' *Mathematical Gazette*, XL (May 1956), 129.
46. Levi Leonard Conant, *The Number Concept* (New York, Macmillan and Company, 1923), p. 70.

4 ALGEBRA

It would be an injustice to pioneers in mathematics to stress modern mathematical ideas with little reference to those who initiated the first and possibly the most difficult steps. Nearly everything useful that was discovered in mathematics before the seventeenth century has either been so greatly simplified that it is now part of every regular school course, or it has long since been absorbed as a detail in some work of greater generality.[1]

The Muslims translated numerous Greek works in mathematics as they did in other fields of science. At the same time, they turned to the East and gathered all that was available in India in the way of science, and particularly in mathematics.[2] In the field of algebra the Muslims soon made original contributions which proved to be the greatest of their distinctive achievements in mathematics.[3]

Definition of Algebra

Algebra is that 'branch of mathematical analysis which reasons about quantities using letters to symbolize them.'[4] Algebra is defined in the *Mathematics Dictionary* as 'a generalization of arithmetic; e.g., the arithmetic facts that $2 + 2 + 2 = 3 \times 2$, $4 + 4 + 4 = 3 \times 4$, etc., are all special cases of the (general) algebraic statement that $x + x + x = 3x$, where x is any number.'[5] The Muslim scholar ibn Khaldun defined algebra as a 'subdivision' of arithmetic. This is a craft in which it is possible to discover the unknown from the known data if there exists a relationship between them.[6]

In the ninth century the Muslim mathematician Al-Khwarizmi wrote his classical work on algebra, *Al-Jabr wa-al-Mugabala*. In this title the word *Al-Jabr* means transposing a quantity from one side of an equation to another, and *Mugabala* signified the simplification of the resulting expressions.[7] Figuratively, *al-jabr* means restoring the balance of an equation by transposing terms.[8] Because of the double title, the explanation given contains a comment upon the second word, *al-maqabala*, as well as the first one. According to David Eugene Smith:

In the 16th century it is found in English as *algiebar* and *almachabel*, and in various other forms but was finally shortened to *algebra*. The words mean restoration and opposition, and one of the clearest explanations of their use is given by Beha Eddin (1600 AD) in his *Kholasat Al-Hisab* (essence of arithmetic): The member which is affected by a minus sign will be increased and the same added to the other member, this being algebra; the homogeneous and equal terms will then be cancelled, this being *Al-Muqabala.*[9]

That is, given $x^2 + 5x + 4 = 4 - 2x + 5x^3$,
Al-Jabr gives $x^2 + 7x + 4 = 4 + 5x^3$,
Al-Muqabala gives $x^2 + 7x = 5x^3$.

Therefore, the best translation for *Hisab Al-Jabr Wa-al-Muqabala* is 'the science of equations.'[10]

The Origin of the Term 'Algebra'

Al-Khwarizmi's text of algebra, entitled *Al-Jabr Wa-Al-Muqabala* (the science of cancellation and reduction) was written in 820 AD.[11] A Latin translation of this text became known in Europe under the title *Al-Jabr.*[12] Thus 'the Arabic word for reduction, *al-Jabr,* became the word *algebra.*'[13]

Al-Khwarizmi

From the eighth to the thirteenth centuries, the center of scientific activity was Arabia. Scientific activity was centered in the Muslim world, especially at the court of the Caliph Al-Ma'mum.[14] It was there that Al-Khwarizmi (825 AD) influenced mathematical thought more than any other medieval writer by finding a system of analysis for solving equations of first- and second-degree in one unknown by both algebraic and geometric means.[15]

The first half of the ninth century is characterized by Sarton in his *Introduction to the History of Science* as 'the time of Al-Khwarizmi,' because he was 'the greatest mathematician of the time, and if one takes all circumstances into account, one of the greatest of all times.'[16] E. Wiedmann has said: 'His works, which are in part important and original, reveal in Al-Khwarizmi a personality of strong scientific genius.'[17]

David Eugene Smith and Louis Charles Karpinski characterized Al-Khwarizmi as:

> . . . the great master of the golden age of Baghdad, one of the first of the Muslim writers to collect the mathematical classics of both the East and the West, preserving them and finally passing them on to the awakening Europe. This man was . . . a man of great learning and one to whom the world is much indebted for its present knowledge of algebra and of arithmetic.[18]

It was Mohammad Khan who stated:

> In the foremost rank of mathematicians of all times stands Al-Khwarizmi. He composed the oldest works on arithmetic and algebra. They were the principal source of mathematical knowledge for centuries to come both in the East and the West. The work on arithmetic first introduced the Hindu numbers to Europe, as the very name algorism signifies; and the work on algebra not only gave the name to this important branch of mathematics in the European world, but contained in addition to the usual analytical solution of linear and quadratic equations (without, of course, the conception of imaginary quantities) graphical solution of typical quadratic equations.[19]

The mathematics that the Muslims inherited from the Greeks made the division of an estate among the children extremely complicated, if not impossible. It was the search for a more accurate, comprehensive, and flexible method that led Al-Khwarizmi to the innovation of algebra.[20] While engaged in astronomical work at Baghdad and Constantinople, he found time to write the algebra which brought him fame.[21] His book, *Al-Kitab al-Mukhtasar fi hisab al-jabr wa-al-Muqabala*, is devoted to finding solutions to practical problems which the Muslims encountered in daily life.[22]

In evolving his algebra, Al-Khwarizmi transformed the number

> from its earlier arithmetical character as a finite magnitude into an element of relation and of infinite possibilities. It can be said that the step from arithmetic to algebra is in essence a step from 'being' to 'becoming' or from the static universe of the Greek to the dynamic ever-living, God-permeated one of the Muslims.[23]

Al-Khwarizmi emphasized that he wrote his algebra book to serve the practical needs of the people concerning matters of inheritance, legacies, partition, lawsuits, and commerce. He dealt with the topic which in Arabic is known as *'ilm al-fara'id*[24] (the science of the legal shares of the natural heirs).[25] Gandz stated in *The Source of Al-Khwarizmi's Algebra*:

> Al-Khwarizmi's algebra is regarded as the foundation and cornerstone of the sciences. In a sense, Al-Khwarizmi is more entitled to be called 'the father of algebra' than Diophantus because Al-Khwarizmi is the first to teach algebra in an elementary form and for its own sake, Diophantus is primarily concerned with the theory of numbers.[26]

In the twelfth century the algebra of Al-Khwarizmi was translated into Latin by Gerhard of Cremona and Robert of Chester.[27] It was used by Western scholars until the sixteenth century.[28] Of the translation by Robert of Chester, Sarton judicially remarked: 'The importance of this particular translation can hardly be exaggerated. It may be said to mark the beginning of European algebra.'[29]

After Al-Khwarizmi, there were many other Muslims who studied and taught algebra, but they made few new discoveries. They were content to know what he had written in his great book.[30]

Roots

The term 'root' had its origin in the Arab language. 'Latin works translated from the Arabic have *radix* for a common term, while those inherited from the Roman civilization have *latus*.' *Radix* (root) is the Arabic *jadhr,* while *latus* referred to the side of a geometric square.[31]

The Arabic word for root was used by Al-Khwarizmi to denote the first-degree term of a quadratic equation. Al-Khwarizmi wrote, 'the following is an example of squares equal to roots: a square is equal to 5 roots. The root of the square then is 5, and 25 forms its square, which of course equals 5 of its roots.'[32]

Square Root

The method of extracting the square root employed by the Muslims resembled their method of division. For example, to find the square root of 107584 vertical lines are drawn and numerals are partitioned into periods of two digits. See details in Figure 4.1. The nearest root of 10 is 3, which is placed both below and above, and its square, 9, is subtracted from 10. The 3 is now doubled and the result is written in the next column. Six is contained twice in 17, the remainder with first figure of the next period. The 2 is set down both above and below, and being multiplied by 6 gives 12, which subtracted from 17, leaves 5. The square of 2, or 4, is now subtracted from 55. The difference 51, together with the succeeding figure, or 518, is divided by the double of 32, or 64, giving 8 for the quotient. Then 8 times 64, or 512, is subtracted from 518 with a difference of 6. This digit together with the succeeding figure forms 64 which is exhausted by subtracting from it the square of 8. Therefore, the square root of 107584 is 328. It has been said that this method was adopted from the Muslims by the Hindus.[33]

Al-Karkhi, another Muslim mathematician, employed a method of approximation to find the square root of numbers using the formula.[34]

$$\sqrt{a} = w + \frac{a - w^2}{2w + 1}$$

The approximate root for $\sqrt{17} = 4 = w$. Therefore

$$\sqrt{17} = 4 + \frac{17 - 16}{2(4) + 1} = 4\frac{1}{9} = 4.1111$$

and this value checks fairly well with 4.123106, the value precise to six figures.

1	0	7	5	8	4
	9				
	1	7			
	1	2			
		5	5		
			4		
		5	1	8	
		5	1	2	
				6	4
				6	4
				0	0
		6	6		
	3		2		

Fig. 4.1: An Illustration of the Arabian Method for Extracting Square Root

Linear and Quadratic Equations

The Egyptians solved equations of the first degree more than four thousand years ago. That is, they found that the solution of the equation $ax + b = 0$ is $x = -b/a$. The graph of the equation is represented in geometry by a straight line. The quadratic equation, however, $ax^2 + bx + c = 0$ was solved by the Muslims with the formula:

$$x = \frac{-b \pm \sqrt{(b^2 - 4ac)}}{2a}$$

The various conic sections, such as the circle, the ellipse, the parabola, and the hyperbola are the geometric representations of quadratic equations in two variables which were studied by the Muslims.[35]

In his work on linear and quadratic equations, Al-Khwarizmi used special technical terms for the various multiples or powers of the unknown. The unknown is referred to as a 'root' and the unknown squared is called the 'power.' With this vocabulary, Al-Khwarizmi would describe general linear equations as 'roots equal to numbers.' In present-day notation it would appear as $bx = c$. Instances of linear equations are: one root equals three, $x = 3$; four roots equal twenty, $4x = 20$; and one-half a root equals ten $(½)x = 10$.[36]

Al-Khwarizmi separated general quadratic equations into five cases for purposes of finding the solution to a given equation. The five cases he considered were: (1) squares equal to roots, $ax^2 = bx$; (2) squares equal to numbers, $ax^2 = b$; (3) squares and roots equal to numbers, $ax^2 + bx = c$; (4) squares and numbers equal to roots, $ax^2 + c = bx$; and (5) squares equal to roots and numbers, $ax^2 = bx + c$. In all applications, Al-Khwarizmi considered a, b, c positive integers with $a = 1$. He was concerned with only positive real roots, but he recognized the existence of a second root which was not conceived of previously.[37] Examples are given of cases (3), (4), and (5) above to illustrate the methods of Al-Khwarizmi.

Case (3): Square and roots equal to numbers, $x^2 + 10x = 39$.

Construct the square ABCD with side AB = x. Extend AD to E and AB to F such that DE = BF = (½) 10 = 5, then complete the square AFKE. By extending DC to G and BC to H, the area of square

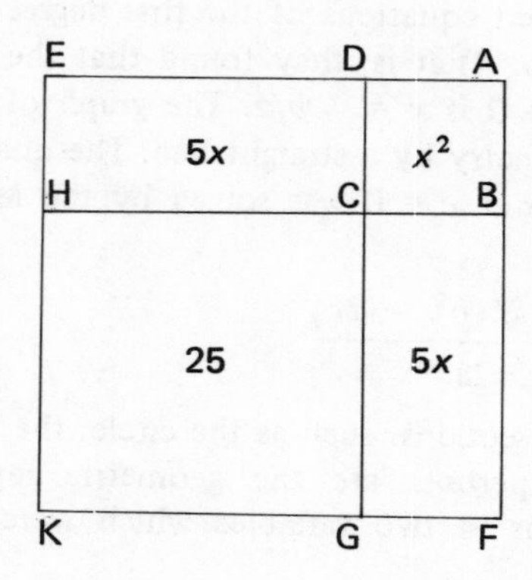

Fig. 4.2: Paradigm of Al-Khwarizmi's for Quadratic Equation
$x^2 + 10x = 39$

AFKE may be expressed as $x^2 + 10x + 25$. However, the equation to be solved is $x^2 + 10x = 39$. Therefore, 25 must be added to each member of this equation to yield $x^2 + 10x + 25 = 39 + 25 = 64$, which is the required area. In other words, $x^2 + 10x + 25$ is a perfect square $(x + 5)^2$, and this is equal to another perfect square, 64. Hence, the dimensions of the area $(x + 5)^2$ must by 8 by 8. However, since AF $= x + 5 = 8$, this means $x = 3$.[38]

Case (4): Squares and numbers equal to roots, $x^2 + 21 = 10x$, $x < b/2$, *where a is coefficient of x*

Construct a rectangle ABCD with side AB $= x$ and BC $= 10$. The area of rectangle ABCD $= 10x = x^2 + 21$. On side BC mark off a point E such that BE $=$ BA, then complete the square ABEF. It follows that the area of rectangle CDFE $= 21$. Let H be the midpoint of BC. Extend side CD to N such that CN $=$ CH $= 5$ and complete the square HCNM, whose area is 25. From I, the midpoint of AD, construct the point S such that IS $=$ IF $= 5 - x$ and complete the square MISW, whose area is $(5 - x)^2$. Since DS $= x$, the area of rectangle DSWN $= x\,(5 - x) =$ the area of rectangle FEHI. Therefore,

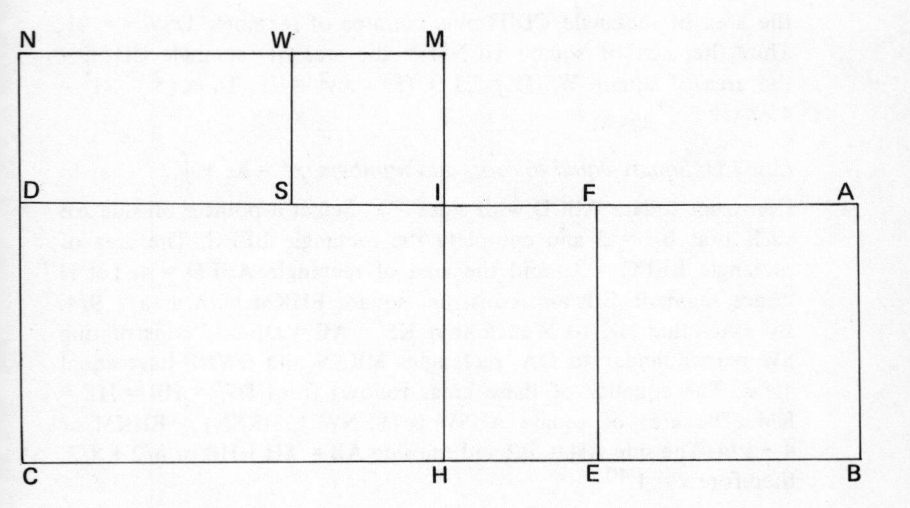

Fig. 4.3: Paradigm of Al-Khwarizmi for Quadratic Equation
$x^2 + 21 = 10x$

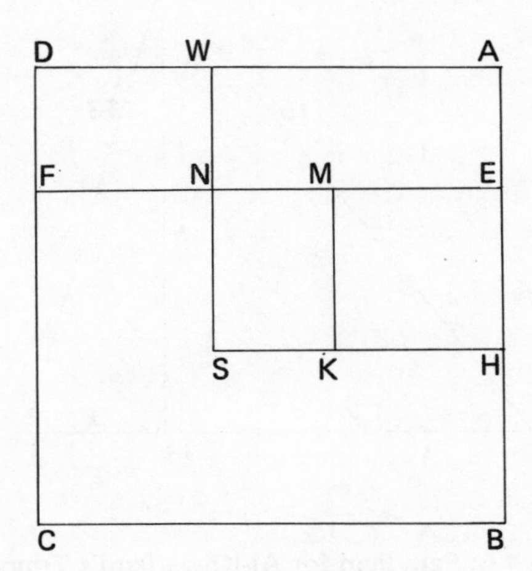

Fig. 4.4: Paradigm of Al-Khwarizmi for Quadratic Equation
$x^2 = 3x + 4$

the area of rectangle CDIH plus the area of rectangle DSWN = 21. Thus the area of square HCNM = the area of rectangle CDFE + the area of square WSIM = 21 + $(5 - x)^2$ = 25. Then $(5 - x)^2$ = 4 or $x = 3$.[39]

Case (5): Square equal to roots and numbers, $x^2 = 3x + 4$

Construct square ABCD with sides = x. Select a point E on side AB such that BE = 3 and complete the rectangle BEFC. The area of rectangle BEFC = $3x$ and the area of rectangle AEFD = 4. Let H bisect segment EB, and construct square EHKM with area = 9/4. By extending HK to S such that KS = AE = DF and constructing SW perpendicular to DA, rectangles MKSN and DWNF have equal areas. The equality of these areas follows from DW = HB = HE = KM. The area of square AHSW is (SENW + MKSN) + EHKM or 4 + 9/4. The side AH = 5/2 and the side AB = AH + HB or 5/2 + 3/2, therefore $x = 4$.[40]

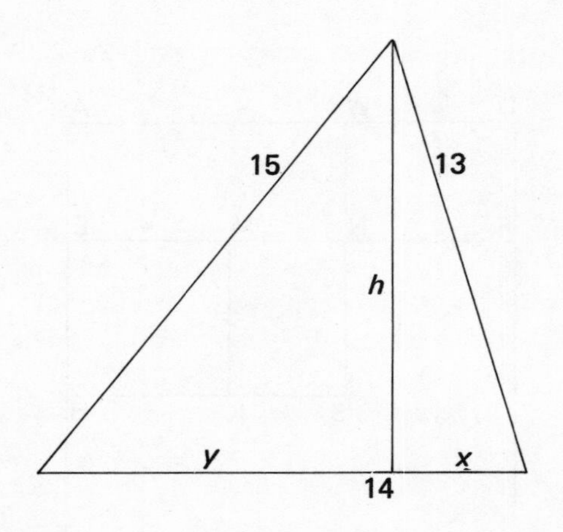

Fig. 4.5: Paradigm for Al-Khwarizmi's Triangle

Al-Khwarizmi gave an algebraic method for finding the altitude and the two segments of the base formed by the foot of the altitude, x and y, of the triangle when the three sides (13, 14, and 15) are given, as in Figure 4.5. The square of the height, h^2, is equal to $13^2 - x^2 = 15^2 - y^2 = 15 - (14 - x)^2$. Hence $169 - x^2 = 225 - 196 + 28x - x^2$. And upon simplification, $x = 5$; and it follows that $h^2 = 169 - 25$ and that $h = 12$.[41]

The change from the Greek conception of a static universe to a new dynamic one was initiated by Al-Khwarizmi who was the herald of modern algebra, and the first mathematician to make algebra an exact science. After dealing with equations of second degree, Al-Khwarizmi discussed algebraic multiplication and division.[42]

Miscellaneous

Following the period of Al-Khwarizmi's works came those of Thabit ibn Qurra (836-901 AD), mathematician and linguist. His chief contribution to mathematics was in his translations of Euclid, Archimedes, Apollonius, and Ptolemy. Fragments of some original writing in the area of algebraic geometry have also been preserved. This particular branch of algebra received considerable attention from Muslim mathematicians.[43] According to Karl Fink:

> Al-Khwarizmi calls a known quantity a number, the unknown quantity *jidr* (root) and its square *mal* (power). In Al-Karkhi we find the expression *kab* (cube) for the third power, and there are formed from these expressions mal mal = x^4, mal kab = x^5, kab kab = x^6, mal mal kab = x^7, etc.[44]

In his *History of Mathematics*, David Eugene Smith stated:

> . . . Al-Haitham of Basra, who wrote on algebra, astronomy, geometry, gnomic, and optics attempted the solution of the cubic equation by the aid of conics . . .[45]

The Muslims discovered the theorem that for integers the sum of two cubes can never be a cube. This theorem was later rediscovered by P. Fermat, a French physician, and claimed by him. Creditable work in number theory and algebra was done by Al-Karkhi of Baghdad, who lived at the beginning of the eleventh century. His treatise on algebra is sometimes considered the greatest algebraic work of Muslim mathematicians; it shows the influence of Diophantus. For the solution of quadratic equations, he gave both arithmetical and geometrical proofs.[46]

Al-Karkhi's work contains basic algebraic theory with application to equations and especially to problems to be solved for positive rational numbers. For instance, to find two numbers the sum of whose cubes is a square number yields the algebraic expression: $x^3 + y^3 = z^2$. To solve the equation in rational numbers, let:

$$y = mx,\ z = nx;\ x^3 + m^3x^3 = n^2x^2;\ x^3(1 + m^3) = n^2x^2$$

By cancellation of x^2, therefore,

$$x = \frac{n^2}{1 + m^3}$$

where m and n are arbitrary positive rational numbers.[47] As a special solution, Al-Karkhi gave the following values: $x = 1, y = 2, z = 3$. The same method is clearly applicable to many more general rational problems having the form $ax^n + by^n = cz^{n-1}$.[48]

One of the oldest methods for approximating the real root of an equation $ax + b = 0$ is often called the rule of double false position. The Muslims called the rule *hisab al-Khataayn;* it is found in the works of Al-Khwarizmi. This rule seems to have come from India, but it was the Muslims who made it known to European scholars. In order to explain the rule, let g_1 and g_2 be any guessing values of x, and let f_1 and f_2 be the errors. Therefore, if the guesses were right, then $ag_1 + b = 0; ag_2 + b = 0$. However, if the guesses were wrong, then

$$ag_1 + b = f_1 \quad (1)$$
$$ag_2 + b = f_2 \quad (2)$$
$$a(g_1 - g_2) = f_1 - f_2 \quad (3)$$ by subtraction of (2) from (1)

From (1)

$$ag_1g_2 + bg_2 = f_1g_2$$

and from (2) by subtraction

$$ag_1g_2 + bg_1 = f_2g_1$$

Therefore

$$b(g_2 - g_1) = f_1g_2 - f_2g_1 \ . \quad (4)$$

Dividing (4) by (3)

$$\frac{b(g_2 - g_1)}{a(g_1 - g_2)} = \frac{f_1g_2 - f_2g_1}{f_1 - f_2}$$

or

$$\frac{-b}{a} = \frac{f_1g_2 - f_2g_1}{f_1 - f_2}$$

But, since

$$\frac{-b}{a} = x \ ,$$

therefore,

$$x = \frac{f_1g_2 - f_2g_1}{f_1 - f_2}$$

Suppose, for example, that $2x - 5 = 0$; guessing value for x:$g_1 = 5$, $g_2 = 1$.

Then, $2 . 5 - 5 = 5 = f_1$ and $2 . 1 - 5 = -3 = f_2$.

But

$$x = \frac{f_1g_2 - f_2g_1}{f_1 - f_2} = \frac{5 . 1 - (-3) . 5}{5 - (-5)} = \frac{20}{8} = 2 . 5$$ [49]

According to Howard Eves, this method was used by the Muslims and can be illustrated geometrically by letting x_1 and x_2 be two numbers lying close to and on each side of a solution x of the equation $f(x) = 0$. The intersection with the x-axis of the chord joining the point $(x_1, f(x_1))$, $(x_1, f(x_2))$ gives a better approximation to the solution:

$$x_3 = \frac{x_2 f(x_1) - x_1 f(x_2)}{f(x_1) - f(x_2)} .$$

The process can now be applied with appropriate pairs x_1, x_3 or x_3, x_2 depending on circumstances.[50] This is the numerical method of 'False Positions' *(Regular Falsi)* which is used in numerical analysis today.

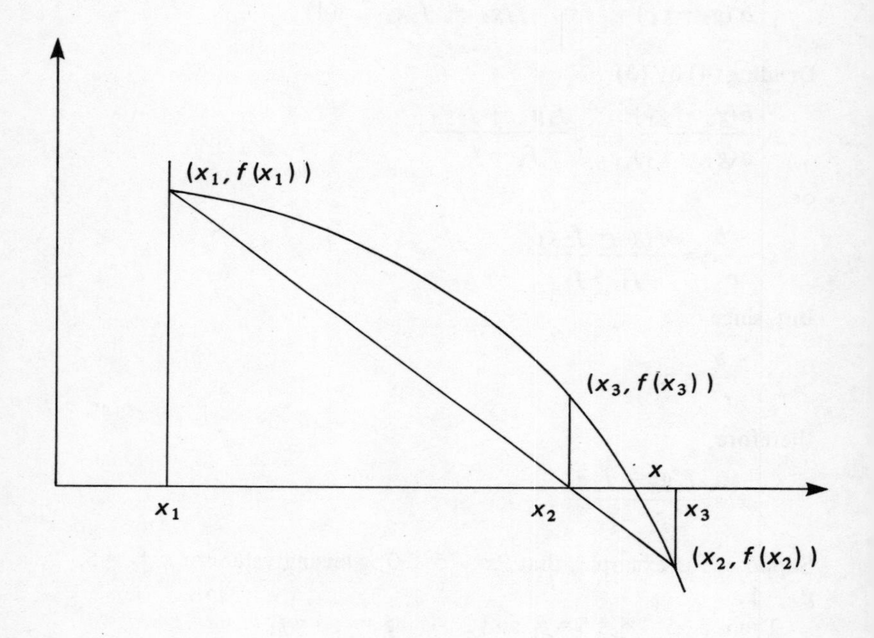

Fig. 4.6: Rule of Double False Position of Al-Khwarizmi

Summary

The Muslims not only created algebra, which was to become the indispensable instrument of scientific analysis, but they laid the foundations for methods in modern experimental research by the use of mathematical models. Since Mohammed ibn Musa Abu

Djefar Al-Khwarizmi was the founder of the Muslim school of mathematics, the subsequent Muslim and early medieval works on algebra were largely founded on his algebraic treatise. Al-Khwarizmi's work plays an important role in the history of mathematics, for it is one of the main sources through which Arabic numerals and Muslim algebra came to Europe.

The contribution of Muslim mathematicians to the field of algebra includes methods for finding the solution to linear and quadratic equations. Solutions to these equations were also given by geometric methods. Al-Karkhi contributed rational solutions to certain special equations of degree higher than two and a method for approximating the solution to linear equations. These are but a few of the more outstanding developments in algebra that resulted directly from the efforts of Muslim mathematicians.

Notes

1. E. T. Bell, *Men of Mathematics* (New York, Simon and Schuster, 1937), p. 14.
2. Daoud S. Kasir, *The Algebra of Omar Khayyam* (New York, J. J. Little and Ives Company, 1931), p. 16.
3. Bodleian Library, Oxford, England, Huntingdon MSS, 214, fol. ff34^{r}-52^{r}.
4. Isaac Funk, Calvin Thomas, and Frank H. Vizetelly (Supervisors), *New Standard Dictionary of the English Language* (New York, Funk and Wagnallis Company, 1940), p. 70.
5. Glenn James and Robert C. James (eds.), *Mathematics Dictionary* (New York, D. Van Nostrand Company, 1963), p. 17.
6. Franz Rosenthal, trans., *The Migaddimah Ibn Khaldun: Autobiography* (New York, Bollingen Foundation, 1958), Vol. III, p. 124.
7. Isma'il Mazhar, *Tarikh al-Fikr al-'Arabi: fi Nushu' ih wa Tatwirih bi Ttarjamah wa Annagil 'an al-Hadarah Al-Yunaniyah* (Cairo, Dar al-'Uswr li Itab' wa Annashur bi Masr).
8. Morris Kline, *Mathematics and the Physical World* (New York, Thomas Y. Crowell Company, 1959), p. 69.
9. David Eugene Smith, *History of Mathematics* (New York, Ginn and Company, 1925), Vol. II, p. 388.
10. John K. Baumgart, 'History of Algebra,' *Historical Topics for the Mathematics Classroom*, Thirty-First Yearbook of the National Council of Teachers of Mathematics (Washington, D.C., National Council of Teachers of Mathematics, 1969), pp. 233-4.
11. Sidney G. Hocker, Wilfred E. Barnes, and Calvin T. Long, *Fundamental Concepts of Arithmetic* (Englewood Cliffs, N.J., Prentice-Hall, 1963), p. 9.

12. H. A. Freebury, *A History of Mathematics: For Secondary Schools* (London, Cassell and Company, 1968), p. 77.
13. Solomon Gandz, 'The Origin of the Term Algebra,' *The American Mathematical Monthly*, XXXIII (May 1926), 437.
14. Edna E. Kramer, *The Nature and Growth of Modern Mathematics* (New York, Hawthorn Books, 1970), p. 85.
15. Franklin W. Kokomoor, *Mathematics in Human Affairs* (New York, Prentice-Hall, 1946), p. 172.
16. George Sarton, *Introduction to the History of Science* (Baltimore, The Williams and Wilkins Company, 1927), Vol. I, p. 563.
17. M. Th. Houtsma, T. W. Arnold, R. Basset, and R. Hartmann (eds.), *The Encyclopaedia of Islam* (London, Luzac and Company, 1913), Vol. I, p. 912.
18. David Eugene Smith and Louis Charles Karpinski, *The Hindu-Arabic Numbers* (Boston and London, Ginn and Company, 1911), pp. 4-5.
19. Mohammad Abdur Rahman Khan, *A Brief Survey of Moslem Contribution to Science and Culture* (Lahore, Sh. Umar Daraz at the Imperial Printing Works, 1946), pp. 11-12.
20. Rom Landau, *The Arab Heritage of Western Civilization* (New York, Arab Information Center, 1962), pp. 33-4.
21. Florence Annie Yeldham, *The Story of Reckoning in the Middle Ages* (London, George G. Harrap and Company, 1926), p. 64.
22. Lancelot Hogben, *Mathematics for the Million* (New York, W. W. Norton and Company, 1946), pp. 290-1.
23. Rom Landau, *Arab Contribution to Civilization* (San Francisco, The American Acadamy of Asian Studies, 1958), p.33.
24. Bodleian Library, Oxford, England, Marsh MSS, 640, fol. (f. 102).
25. Solomon Gandz, 'The Algebra of Inheritance,' *Osiris*, V (1938), 324.
26. Solomon Gandz, 'The Source of Al-Khwarizmi's Algebra,' *Osiris* (Bruges, Belgium, The Saint Catherine Press Ltd., 1936), Vol. I, p. 264.
27. Ibid., p. 263.
28. Joseph Hell, *The Arab Civilization* (Lahore, Sh. Mohd. Ahmad at the Northern Army Press, 1943), p. 95.
29. George Sarton, *Introduction to the History of Science* (Baltimore, The Williams and Silkins Company, 1953), Vol. II, Part I, p. 176.
30. Walter H. Carnahan, 'History of Algebra,' *School Science and Mathematics*, XLVI, 399 (January 1946), 10.
31. Solomon Gandz, 'Arabic Numerals,' *American Mathematical Monthly*, XXXIII (January 1926), 261.
32. Philip S. Jones, ' "Large" Roman Numerals,' *The Mathematics Teacher*, CXVII (March 1954), 196.
33. Indian Office Library, London, England, Arabic MSS, 757, fol. 4^{b}-5^{a}.
34. George E. Reves, 'Outline of the History of Algebra,' *School Science and Mathematics*, III (January 1952), 63.
35. Edward Kasner and James Newman, *Mathematics and the Imagination* (New York, American Book Stanford Press, 1945), p. 17.
36. Ali Mustafa Mashrafah Wa Mohammed Musa Ahmad (ed.), *Kitab Al-Jabr wa-Al-Muqabala Li Mohammed Ibn Musa Al-Khwarizmi* (Cairo, Dae Al-Katib Al-Arabi Littibah wa Al-Nashr, 1968), p. 16.
37. Aydin Sayili, *'Abd al-Hamid ibn Turk and the Algebra of His Time* (Ankara, Turk Tarih Kuruma Basimeni, 1962), p. 146.

38. Walter H. Carnahan, 'Geometric Solutions of Quadratic Equations,' *School Science and Mathematics*, XLVII, No. 415 (November 1947), pp. 689-90.
39. Martin Levey, *The Algebra of Abu Kamil* (Madison, The University of Wisconsin Press, 1966), pp. 23-4.
40. Louis Charles Karpinski, trans., *Robert of Chester's Latin Translation of Algebra of Al-Khwarizmi* (London, Macmillan and Company, 1915), p. 87.
41. Henrietta O. Midonick (ed.), *The Treasure of Mathematics* (New York, Philosophical Library, 1965), pp. 432-3.
42. Landau, op. cit., pp. 31-2.
43. Lynn Thorndyke, *A Short History of Civilization* (New York, F. S. Crofts and Company, 1930), p. 292.
44. Karl Fink, *A Brief History of Mathematics* (The Open Court Publishing Company, 1900), p. 75.
45. David Eugene Smith, *History of Mathematics* (New York, Dover Publications, 1958), Vol. I, pp. 175-6.
46. Florian Cajori, *A History of Mathematics* (New York, The Macmillan Company, 1924), p. 106.
47. Howard Eves, *An Introduction to the History of Mathematics* (New York, Holt, Rinehart and Winston, 1969), 3rd edn., p. 201.
48. Oystein Ore, *Number Theory and Its History* (New York, McGraw-Hill Book Vompany, 1948), pp. 185-7.
49. Smith, *History of Mathematics,* op. cit., Vol. II, pp. 437-9.
50. Eves, op. cit., p. 203.

5 TRIGONOMETRY

The Muslim mathematicians' great interest in arithmetic, number theory, and algebra appears to have led them into related areas of the applied and theoretical sciences. They had more than passing interest in the works of earlier civilizations. One can see something of this interest in the fact that they translated virtually all known information of their day into Arabic. The scholarly endeavors of search, translation, and research were in evidence in many of the more important centers of Muslim learning.

Arithmetic, and its application to the commercial and business needs of Muslim life, became the impetus for further study and research into mathematics. It seems natural that the Muslims should turn early to an investigation of the fields of astrology and astronomy. Here, in addition to arithmetic, they needed the use of trigonometry to present a clear model of the heavens and its relationships to their mode of life.

Trigonometry, the handmaiden of astronomy, became an absorbing study for Muslim mathematicians, and it led to several useful and well-known studies in the sciences. For example, the Muslims appear to have been the first to give serious study to the principles of light. Al-Haitham wrote an important treatise on optics which became a classic for several centuries. In his treatise, Al-Haitham set forth the early form of what was to become Snell's law of refraction of light. Al-Haitham's *Optics* inspired attention to astronomy and trigonometry and formed the basis of scholarly research during the Dark Ages and Middle Ages. There is little doubt that these investigations provided considerable support for such men as Leonardo da Vinci, Galileo, and Newton.

The initial development of trigonometry was evidently motivated by a need for numerical solutions to problems related to spherical astronomy. Its emergence as a branch of mathematics, independent of astronomy, was surely a slow process.[1] Perhaps more than any other branch of mathematics, trigonometry appears to have developed as a result of a continual interplay between the supply of applicable mathematical theories and the demands of the science of astronomy.[2] Their relationship was considered intimate until the Renaissance (1450-1600 AD), during which time trigonometry was treated as an auxiliary topic to astronomy, while the problems of mathematical astronomy were related to spherical trigonometry.[3]

After the decline of Alexandria, Greek science lingered only in southern Italy and in Byzantium but was later revived and spread by the Muslims. In the ninth and tenth centuries, while Europe was still in relative darkness, Muslim science and culture were at their height.[4] An investigation of the development of trigonometry during the twelfth and thirteenth centuries will serve as an indication of the progressive work in science conducted by the Muslims.[5]

Definition

The fundamental idea of trigonometry is the measurement of distances indirectly. It would be a physical impossibility to measure directly the height of the great pyramid in Egypt or other inaccessible distances, such as the width of a gorge to be bridged. These and many other problems in the field of surveying and navigation depend upon the solution of triangles.[6]

The word trigonometry comes from the Greek words *tri*, meaning 'three'; *gonon*, 'angle'; and *metria*, 'measurement.' These terms explain the primary significance of this branch of mathematics.[7] George Howe defined trigonometry as 'the science of angles; its province it to teach how to measure and employ angles with the same ease that we handle lengths and areas.'[8] Trigonometry has been defined also as the measurement and calculations of the sides and angles of a triangle.[9]

Although the term *trigonometry* was not used until 1595, when it was initially introduced by Pitiscus,[10] the Muslims had worked diligently and meticulously on the early development of the science.[11] The following sections will present the Muslim contributions to trigonometry.

The Origin of Trigonometry

Trigonometry as it is known today as a branch of mathematics that is linked with algebra. As such, it is to be dated back to the eighth century. When treated purely as a development of geometry, however, it is dated to the time of the great Greek mathematicians and astronomers who flourished about two hundred years before and after the beginning of the Christian era. If regarded simply as 'tri-angle-measurement,' which is all the word trigonometry implies, its roots go back to the Egyptian period four thousand years ago.[12]

By the middle of the twelfth century, Latin mathematicians were acquainted with Muslim trigonometry, if not in its very latest state, at least in the one it had reached before the end of the preceding century.[13] Practically all of the advanced trigonometrical work of the twelfth and thirteenth centuries was produced by Muslim mathematicians and written in the Arabic language. Latin trigonometry was a pale reflection of Muslim trigonometry at this time,[14] and it would wait until the fourteenth century before gaining importance at Merton College in Oxford.[15]

The trigonometry of Muslims is based on Ptolemy's theorem but is superior in two important respects. It employs the sine where Ptolemy used the chord and is in algebraic instead of geometric form.[16] The theory of sine, cosine, and tangent is a legacy of the Muslims. The brilliant epochs of Peurbach, Regiomontanus, and Copernicus cannot be recalled without being reminded of the fundamental and prepartory labors of the Muslim mathematicians.[17]

Consider the movement of a line (now known as the radius vector) in a counterclockwise direction round a fixed point. The perpendiculars drawn from the end point of this line, in its various positions, to the original direction form segments which correspond to the half-chords referred to by Ptolemy. The length of these segments or half-chords became associated with an angle through which the revolving line turned.[18]

The half-chord in Arabic is known as *jiba* and became confused with *jaib*. Arabic words were frequently written without vowels, and the consonants of both *jiba* and *jaib* are *j* and *b*.[19] *Jaib*, however, had nothing to do with the length of a half-chord since it meant 'the opening of a garment at the neck and bosom.' By the time European mathematicians became familiar with Arabic terms concerning half-chords, they were consistently calling half-chords by the term *jaib* with its meaningless reference to the bosom. Consequently, the European mathematicians translated *jaib* by the Latin *sinus*, meaning 'bosom' or 'fold.' The term *sine* for the half-chord is derived from the Latin *sinus.*[20]

While trigonometry was at first treated as a branch of astronomy, it was eventually studied independently. The Muslims were vastly superior to the Greeks and the Indians in the area of trigonometry in so far as they enlarged or used tables of the six fundamental trigonometrical functions and established the funamental relations between them.[21]

Al-Battani

In the ninth century, just as he does today, man studied the mystery of God and the relationship between heaven and earth. Therefore, it is not surprising that the Muslims directed their attention to spherical trigonometry, and Al-Battani became their chief proponent.[22] Mohammed ibn Jabir ibn Sinan Abu Abu Abdullah Al-Battani was born in Battan, Mesopotamia in 850 and died in Damascus in 929 AD.[23] He was an Arabian prince, governor of Syria, and is considered the greatest Muslim astronomer and mathematician.[24]

Al-Battani is mainly responsible for the modern concepts and notations of trigonometrical functions and identities.[25] Various astrological writings, including a commentary on Ptolemy's *Tetrabiblon* ('four books'), are attributed to him, but his main work was an astronomical treatise with tables, *De Scientia* and *De Numeris Stellarum et Motibus* (About Science and Number of Stars and their Motion) which was extremely influential until the Renaissance. He made astronomical observations of remarkable range and accuracy throughout his life. His tables contain a catalog of fixed stars compiled during the year 800-1. He found that the longitude of the sun's apogee had increased 16 degrees and 47 minutes since Ptolemy's planetary theory, 150 AD. That implies the discovery of the motion of the solar apsides. Al-Battani determined several astronomical coefficients with great accuracy; precession, 5415 seconds a year; inclination of the eliptic, 23 degrees and 35 minutes. He proved the possibility of annual eclipses of the sun, and he did not believe in the trepidation of the equinoxes.[26]

At that time, Al-Battani's work in astronomy was devoted mainly to trigonometry. He used sines regularly with a clear consciousness of their superiority over the Greek chords. Al-Battani, who was called 'Albategnius' by the Latins, completed the introduction of functions *umbra extensa* and *umbra versa* (cotangents and tangents) and gave a table of cotangents in terms of degrees.[27] He also knew the relation between the sides and angles for the general spherical triangle, which is expressed by the formula $\cos a = \cos b \cos c + \sin b \sin c \cos \alpha$, see Figure 5.1-A.[28] For the spherical right triangle with the right angle at C, he gave the formula, $\cos \beta - \cos b \sin \alpha$, see Figure 5.1-B.[29]

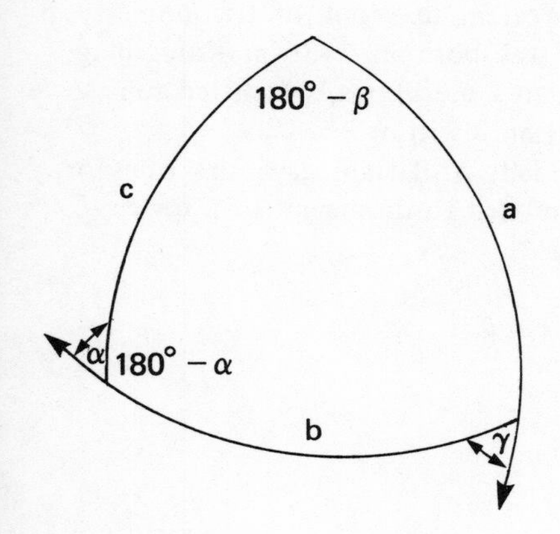

Figure 5.1-A

Figure 5.1-B

Al Battani not only computed sine, tangent, and cotangent tables from zero to 90 degrees with great accuracy, but he also applied algebraic operations to the trigonometric identity for the spherical triangle.[30] He developed the tables of cotangents based on the relation $\cot \alpha = \cos \alpha / \sin \alpha$.[31]

Al Battani also wrote such books as 'The book of the science of the ascensions of the signs of the zodiac in the spaces between the quadrants of the celestial sphere,' 'A letter on the exact determination of the quantities of the astrological applications,' and 'Commentary on Ptolemy's Tetrabiblon.' His principal work was *al-Zij* ('Astronomical Treatise and Tables'). This book contains the results of his observations and had a considerable influence not only on astronomy in the Muslim world but also on the development of astronomy and spherical trigonometry in Europe during the Middle Ages and the beginning of the Renaissance.[32]

Al-Battani wrote about the tangent, but its advantages apparently were not recognized by early Western scientists. In the thirteenth century many mathematicians referred to it as the *umbra,* and in the fourteenth century Levi Ben Gershon discussed the tangent in his *De sinibus chords, et arcubus, item instrumento revelatore secretorum* (about the sine, chords and arcs, also about instruments as yet being secrets), the first Western textbook of trigonometry. However, Regiomontanus, who was born in 1436 in Konigsberg, was one to appreciate the tangent's usefulness, which led him to revise the entire book after Al-Battani's writing.[33]

Called the Ptolemy of Baghdad, Al-Battani gave the rule for finding the altitude of the sun related to the height of a tower, L, and its shadow, x, by the formula:[34]

$$x = \frac{L \sin (90 - \theta)}{\sin \theta} = L \cot \theta \ .$$

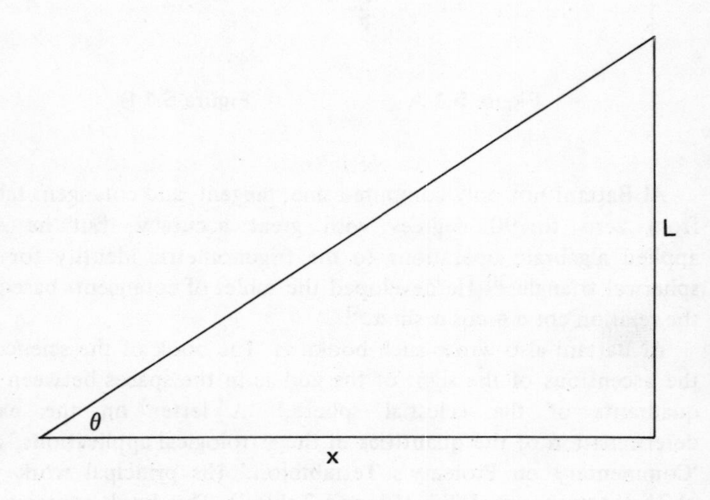

Figure 5.2: Al-Battani's Plane Triangle

In discovering the motion of the sun's apogee, Al-Battani showed Ptolemy's error (of 17 degrees). By calculating the length of the year to be 365 days, 5 hours, 46 minutes, and 24 seconds, he was accurate to within two minutes of the exact time. He also corrected other observations of Ptolemy by establishing tables relevant to the motion of the sun, moon, and planets.[35]

Other Famous Muslim Mathematicians

The works of other Muslim mathematicians, Al-Biruni, Ibn Al-Shatir, Al-Khwarizmi, and Ibn Al-Haitham, also contributed to the development of trigonometry. It will be noted that many of these mathematicians also made major developments in other areas of knowledge.

Al-Biruni

Al-Biruni was among those who laid the foundation for modern trigonometry.[36] As a philosopher, geographer, and astronomer,[37] Al-Biruni was not only a mathematician but a physicist as well. His contribution to physics was through studies in specific gravity and the origin of artesian wells.[38] Al-Biruni resided in India for nearly thirteen years (1017-30) and devoted himself to the study of the arts and sciences of the Hindus. He also had a remarkable knowledge of the Greek sciences and literature.[39]

Taki Ed Din al-Hilali considers Al-Biruni to be 'one of the very greatest scientists of all time.'[40] Six hundred years before Galileo, Al-Biruni had discussed the possibility of the earth's rotation around its own axis.[41]

Al-Biruni carried out geodesic measurements[42] and determined the magnitude of the earth's circumference in a most ingenious manner.[43] With the aid of mathematics, he fixed the direction to Mecca in mosques all over the world.[44]

Ibm Al-Shatir

'Ala al-Din 'Ali ibn Ibrahim Ibn Al-Shatir Al-Muwaqqit, an Arabian trigonometrist, lived from March 1306 until 1375 AD; he died in Damascus. His work progressed in Damascus where he was Muezzin at the great mosque Jami' al-Umawi.

Ibn Al-Shatir is considered one of the outstanding astronomers of his time. He made valuable astronomical observations, and wrote a special treatise, *Rasd Ibn Shatir* (Observatory of Ibn Shatir), concerning them. He devised astronomical instruments and wrote various treatises explaining their structure and use. With regular and precise observations, Ibn Al-Shatir investigated the motion of the celestial bodies and determined at Damascus the obliquity of the ecliptic to be 23 degrees 31 minutes in 1365; the correct value extrapolated from the present one is 23 degrees 31 minutes and 19.8 seconds.[45]

In Ibn Al-Shatir's book, a text of final inquiry in amending the elements, the Ptolemaic eccentric deferent was dispensed with completely and a second epicycle was introduced. Both the solar and lunar models are non-Ptolemaic, and what is of greatest interest is that the lunar theory is identical with that of Copernicus (1473-1543 AD) except for trivial differences in parameters.

Ptolemy assumed a circular path for the sun, but the orbit of Ibn Al-Shatir's sun deviated slightly from a circular motion. The major fault of the Ptolemaic lunar model is its exaggeration of the variation in lunar distance. The major Copernican contribution to the lunar theory consisted in the elimination of this Ptolemaic fault.[46]

There is no trace of the heliocentric concept in the treatise of Ibn Al-Shatir. Al-Shatir and Copernicus were compatible in their idea of utilizing only those celestial motions constructible by combinations of uniform circular motions.[47]

There is much similarity between the models of Ibn al-Shatir and those of Copernicus, both systems composed of constant-length vectors rotating at a constant angular velocity. These astronomers abandoned the Ptolemaic equality; the lengths of corresponding vectors in the two systems are however, nearly equal, and are in many cases even identical.[48]

Al-Khwarizmi

The Caliph Al-Ma'mun built an observatory in Baghdad and another on the plains of Tadmor. His patronage stimulated astronomical observations of every kind. Tables of planetary motions were compiled, obliquity of the ecliptic was determined, and geodesic measurements were carefully made. Al-Khwarizmi was one of the first to compute astronomical and trigonometrical tables.[49] Included in Al-Khwarizmi's work in trigonometry were his one hundred tables of sines and cotangents values.[50]

Ibn Al-Haitham

Abu-'Ali al-Hasan ibn Al-Hasan ibn al-Haitham was born in Basrah, Iraq, in 965 AD and died in Cairo in 1039. He was one of the most important Muslim mathematicians and one of the greatest investigators of optics of all times. As a physician, he wrote commentaries on Aristotle and Galen.[51] His fame came from his treatise on optics which became known to Kepler during the seventeenth century.[52] This masterpiece, *Kitab al-Manazir* (Book of Mirrors), had a great influence on the training of later scientists in Western Europe.[53]

Ibn Al-Haitham's writings reveal his precise development of the experimental facilities. His tables of corresponding angles of incidence and refraction of light passing from one medium to another show how he nearly discovered the law of the ratio of sines for any given pair of media, later attributed to Snell. He investigated twilight relating it to atmospheric refraction by estimating the sun's height to be 19 degrees below the horizon at the commencement of the phenomenon in the mornings or at its termination in the evenings. The figure generally accepted now is 18 degrees.[54]

Ibn Al-Haitham estimated the height of the homogeneous atmosphere on this basis to be about 55 miles, a rather close approximation. He understood the laws of the formation of images in spherical and parabolic mirrors. He was also familar with the reasons for spherical aberration and of magnification produced by lenses. He gave a much more sound theory of vision than that of the Greeks, regarding the lens system of the eye itself to be the sensitive part. Ibn Al-Haitham was also able to solve a number of advanced questions in geometrical optics; for example, he solved the case of an aplanatic surface for reflection through his mastery of mathematics.[55] During his later years, Ibn Al-Haitham went to Egypt where he attempted to regulate the course of the Nile River. After working on this project for a time he earned his living by writing mathematical books.[56]

Summary

Trigonometry is a science satisfying two practical requirements.

It inherits from both astronomy (science of celestial bodies) and from geometry (science of earth measurement) its main problem, namely that of measuring an inaccessible distance. According to Edward J. Byng:

> Trigonometry is mainly the original creation of the Arabians. So is analytical geometry, and algebra, whose very name is Arabic. The Arabs solved cubic equations by geometric construction. Following their revolutionary achievements in trigonometry, they invented celestial navigation, still the basis of a modern naval officer's training. Even 'our' terms, as used in modern navigation . . . 'azimuth, zenith, nadir' . . . are Arabic. The magnetic needle was discovered in China, but it was the Arabs who adopted it for navigation, inventing the mariners' compass. They also invented the astrolabe.[57]*

Trigonometry depends upon mathematics, and equally vital are the related instruments for navigation. In that field, too, the Muslims proved to be the chief pioneers. During the Middle Ages, there were no telescopes, electrical gadgets, or radar; and measurements had to be made with purely mechanical instruments, such as the quadrant and the astrolabe. To reduce the margin of error, the Muslims made their instruments larger than any known previously. The most famous observatory at which these instruments were being used was at Maragha, in the thirteenth century, where distinguished astronomers from many countries collaborated.

The Muslims were also acquainted with the elements of spherical trigonometry, associated with Al-Battani, the earliest of the many distinguished Muslim astronomers.

The invaluable contributions of the Muslim mathematicians to trigonometry overshadow their work in the field of geometry. Although they did not extend the theory of geometry, the Muslims did establish a close relationship between geometry and algebra in their geometrical solutions of algebraic problems. Chief among their contributions was the translation of Euclid's *Elements* from Greek to Arabic.

---

* Astrolabe is the predecessor of the modern sextant.

Notes

1. Abbas El-Azzawi, *History of Astronomy in Iraq* (Baghdad, Iraq Academy Press, 1959), p. 17.
2. Edward S. Kennedy, 'The History of Trigonometry,' *Historical Topics for the Mathematics Classroom*, Thirty-first Yearbook National Council of Teachers of Mathematics (Washington, D.C., National Council of Teachers of Mathematics, 1969), p. 333.
3. George Sarton, *The Appreciation of Ancient and Medieval Science During the Renaissance* (1450-1600) (Philadelphia, University of Pennsylvania Press, 1955), p. 160.
4. H. T. Pledge, *Science Since 1450: A Short History of Mathematics, Physics, Chemistry, and Biology* (New York, Philosophical Library, 1947), p. 11.
5. George Sarton, *Introduction to the History of Science: From Rabbi Ben Ezra to Roger Bacon* (Baltimore, The Williams and Wilkins Company, 1953), Vol. II, Part I, p. 11.
6. Lee Emerson Boyer, *Mathematics: A Historical Development* (New York, Henry Holt and Company, 1954), p. 415.
7. A. Hooper, *The River Mathematics* (New York, Henry Holt and Company, 1945), p. 222.
8. George Howe, *Mathematics for the Practical Man* (New York, D. Van Nostrand Company, 1957), p. 81.
9. Charles Hutton, *A Course of Mathematics* (Glasgow, Richard Griffin and Company, 1833), p. 415.
10. Alfred Hooper, *Makers of Mathematics* (New York, Random House, 1948), p. 107.
11. Josper O. Hassler and Rolland R. Smith, *The Teaching of Secondary Mathematics* (New York, The Macmillan Company, 1935), pp. 87-8.
12. The Faculties of University of Chicago, Editorial Advisors, *Encyclopaedia Britannica* (Chicago, Encyclopaedia Britannica, 1969), Vol. 22, pp. 235-6.
13. Sarton, op. cit., Vol. II, Part I, p. 11.
14. Rom Landau, *Arab Contribution to Civilization* (San Francisco, The American Academy of Asian Studies, 1958), pp. 35-6.
15. Sarton, op. cit., Vol. II, Part I, p. 11.
16. Henry B. Fine, *Number System of Algebra* (New York, D. C. Heath and Company, 1890), p. 110.
17. Joseph Hell, *The Arab Civilization* (Lahore, Sh. Mohd. Ahmad, 1943), p. 96.
18. Dirk J. Struik, *A Concise History of Mathematics* (New York, Dover Publications, 1967), p. 74.
19. Indian Office Library, London, England, Arabic MSS, 772, fol. 17[b]-18[a].
20. Hooper, *The River Mathematics*, op. cit., pp. 224-5.
21. Rene Taton, *History of Science: Ancient and Medieval Science from the Beginning to 1450* (New York, Basic Books, 1963), Vol. I, pp. 410-11.
22. F. W. Kokomoor, 'The Status of Mathematics in India and Arabia During the "Dark Ages" of Europe,' *The Mathematics Teacher*, XXIX (January 1936), 229.

23. Abi 'Abdullah Mohammed bin Sinan bin Jabir Al-Haruni, *Kitab Assij Assabi'* (Rome, Tubi 'a bi Madinat Rumiyah al-'Uzma, 1899), p. xi.
24. Stephan and Nandy Ronart, *Concise Encyclopaedia of Arabic Civilization: The Arab East* (New York, Frederick A. Praeger, 1960), p. xi.
25. Kokomoor, op. cit.
26. Bodleian Library, Oxford, England, Arabic MSS, 119, fol. (ff. 49^r-54^r).
27. George Sarton, *Introduction to the History of Science: From Homer to Omar Khayyam* (Baltimore, The Williams and Wilkins Company, 1953), Vol. I, pp. 602-3.
28. J. F. Scott, *A History of Mathematics: From Antiquity to the Beginning of the Nineteenth Century* (London, Taylor and Francis Ltd, 1969), p. 52.
29. Howard Eves, *An Introduction to the History of Mathematics* (New York, Holt, Rinehart, and Winston, 1969), p. 194.
30. George E. Reves, 'Outline of the History of Trigonometry,' *School Science and Mathematics*, LIII, No. 2 (February, 1953), p. 141.
31. Carra De Vaux, 'Astronomy and Mathematics,' *The Legacy of Islam* (London, Oxford University Press, 1931), p. 389.
32. H. A. R. Gibb, J. H. Kramers, E. Levi-Provencal, and J. Schacht (eds.), *The Encyclopaedia of Islam* (London, Luzac and Company, 1960), New Edition, Vol. I, pp. 1104-5.
33. Rene Taton, *History of Science: The Beginnings of Modern Science* (New York, Basic Books, 1964), Vol. II, p. 17.
34. David Eugene Smith, *History of Mathematics* (New York, Ginn and Company, 1925), Vol. II, p. 608.
35. J. Villin Marmery, *Progress of Science* (London, Chapman and Hall, 1895), pp. 33-4.
36. Mohammed Saffauri and Adnan Ifram (trans.), *Al-Biruni on Transits* (Beirut, American University of Beirut Press, 1959), p. 17.
37. Sir William Cecil Dampier, *A Shorter History of Science* (New York, The Macmillan Company, 1945), p. 39.
38. Carl B. Boyer, *A History of Mathematics* (New York, John Wiley and Sons, 1968), pp. 263-4.
39. Bibhutibhusan Datta and Avadhesh Narayan Singh, *History of Mathematics* (Lahore, Motilal Banarsi Das, 1935), Part I, p. 98.
40. Taki Ed Din al-Hilali, *Die Einleitung Zu al-Birunis Steinbuch* (Leipzig, Otto Harrassowitz, 1941), p. vii.
41. Rom Landau, *The Arab Heritage of Western Civilization* (New York, Arab Information Center, 1962), p. 33.
42. Sir William Cecil Dampier, *History of Science* (New York, The Macmillan Company, 1943), p. 82.
43. C. Edward Sachau, *Chronologic Orientalischer Volker, Van al-Beruni* (Leipzig, In Commission Bei F. A. Brockhaus, 1878), pp. 184-6.
44. Jamil Ali (trans.), *Tahdid al-Amakin by al-Biruni* (Beirut, The American University Press, 1966), p. 8.
45. George Sarton, *Introduction to the History of Science* (Baltimore, The Williams and Wilkins Company, 1948), Vol. III, Part II, p. 1524.
46. Victor Roberts, 'The Solar and Lunar Theory of Ibm Al-Shatir, A Pre-Copernican Model,' *Isis*, XLVIII, Part 4, No. 154 (December 1957), p. 428.

47. E. S. Kennedy and Victor Roberts, 'The Planetary Theory of Ibn al-Shatir,' *Isis*, L, Part 3, No. 161 (September 1959), 233.
48. Faud Abbud, 'The Planetary Theory of al-Shatir: Reduction of the Geometric Models to Numerical Tables,' *Isis*, LIII, Part 4, No. 174 (December 1962), 492.
49. Sarton, *Introduction to the History of Science*, op. cit., Vol. I, p. 545.
50. Raymond Clare Archibald, 'Hindu, Arabic, and Persian Mathematics – 600 to 1200,' *American Mathematics Monthly*, LVI (January 1949), 30.
51. Theodore F. Van Agenen, *Beacon Lights of Science* (New York, Thomas T. Crowell Company, 1924), pp. 45-6.
52. Solomon Bochner, *The Role of Mathematics in the Rise of Science* (Princeton, New Jersey, Princeton University Press, 1966), p. 304.
53. Van Wagnen, op. cit.
54. Nagula Shahin, 'Al-Daw'u al-Mustagtabu wa al-Tswiru al-Mghari al-Mulwan,' *Gafilh Azzit*, XX (March-April 1972), pp. 7-8.
55. Ibid.
56. M. Th. Houtsma, A. J. Wensinck, T. W. Arnold, W. Heffening, and E. Levi-Provencal (eds.), *The Encyclopaedia of Islam* (London, Luzac and Company, 1927), Vol. II, p. 382.

6 GEOMETRY

Mathematics had its origin during the earliest events in human history, from the time man found it necessary to count and to measure. These early activities stimulated the eventual development of independent subjects, arithmetic and geometry. Consequently, mathematics has had a dual foundation and two main themes. Arithmetical procedures, counting and measurement, appear to have developed simultaneously with the passage of time.[1]

Modern civilization is positively based on science and technology; modern science being a continuation of an ancient endeavor. Modern civilization could not exist without scientific thought.[2] Euclid's work on geometry entitled *Book of Basic Principles and Pillars* was the first Greek work to be translated for students in Muslim lands.[3]

Translations of various works began under Al-Mansur and were further developed under his grandson, Al-Ma'mun. A prince with a fine intellect, a scholar, philosopher, and theologian, Al-Ma'mun was instrumental in the discovery and translation of the works of ancient people. During the reign of Harun Al-Rashid, Al-Hajjaj ibn Yusuf translated into Arabic several Greek works. Among these translations were the first six books of Euclid and the *Almagest*.[4]* The *Almagest,* written by Claudius Ptolemy of Alexandria, was the most outstanding ancient Greek work on astronomy.[5]

The rationale for acquiring a knowledge of geometry, as regarded by the Muslim mathematicians, is set forth in the writings of Ibn Khaldun:

> It should be known that geometry enlightens the intellect and sets one's mind right. All its proofs are very clear and orderly. It is hardly possible for errors to enter into geometrical reasoning, because it is well arranged and orderly. Thus, the mind that constantly applies itself to geometry is not likely to fall into error. In this convenient way, the person who knows

* The name 'Almagest' is a Latinized version of the Arabic title *Almagesti.*

> geometry acquires intelligence. The following statement was written upon Plato's door: 'No one who is not a geometrician may enter our house.'[6]

The work of the Muslims in the application of geometry to the solution of algebraic equations suggests they were the first to establish the close interrelation of algebra and geometry. This was a leading contribution toward the later development of analytic geometry.[7] The Muslims helped to advance mathematical thought during the Dark Ages. It was during the ninth and tenth centuries that they gave to Europe its first information about Euclid's *Elements.*[8]

Definition of Geometry

Geometry is a science[9] which not only leads to the study of the properties of space,[10] but also deals with the measurement of magnitude.[11] It has as its objective the measurement of extension which has length, width, and height as its three dimensions.[12] The word itself came originally from two Greek words, *geo* meaning earth, and *metre*, measurement. It, therefore, meant the same as the word surveying,[13] which is derived from the Old French, meaning 'to measure the earth.'[14]

The Muslims used an interesting folk-etymology to explain the name of Euclid which was used in form *Uclides*; this word was thought to the composed of *Ucli*, meaning a key, and *Dis*, meaning a measure. When combined, the name meant the 'key of geometry.'[15] Euclid's name has since then remained a synonym for geometry.[16] According to William David Reeve:

> Geometry came to be used to designate that part of mathematics dealing with points, lines, surfaces, and solids or with some combination of these geometric magnitudes.[17]

Origin of Geometry

The first geometrical considerations of mankind are ancient and seem to have their origin in simple observations, beginning from human ability to recognize physical forms by comparing shapes and sizes.[18] There were innumerable circumstances in the life of primitive man that would lead to a certain amount of subconscious geometric discovery. Distance was one of the first geometrical concepts to be developed, and the estimation of the time needed to make a journey led to the belief that a straight line constituted the shortest path from one point to another. It is apparent that even animals seem to realize this instinctively. The need to measure land led to the idea of simple geometric figures, such as, rectangles, squares, and triangles. When fencing a piece of land, the corners were marked first and then joined by straight lines. Other simple geometrical concepts, vertical, parallel, and perpendicular lines, would have originated through practical construction of walls and dwellings.[19]

According to the Greek historian Herodotus (*c.* 450 BC),[20] geometry originated in Egypt because the mensuration of land and the fixing of boundaries were necessitated by repeated inundations of the Nile.[21] An ancient manuscript of the Egyptians named 'Papyrus Rhind,' now in the British Museum in London, and written by Ahmes, a scribe of about 2000 BC, contains rules and formulas for finding areas of fields and capacities of wheat warehouses.[22] During the period of its origin, about 1350 BC, geometry was used largely as a means to measure plane figures and volumes of simple solids.[23] The Egyptian mathematicians excelled in the field of geometry and were equal to the Babylonians.[24] As a deductive science, geometry was started by Thales of Miletas (*c.* 600 BC),[25] who also introduced Egyptian geometry to Greece.[26]

Ibn Al-Haitham

Aristotle and Ibn Khaldun both considered optics as a branch of geometry. Progress made in the field of optics would certainly have been impossible in medieval times without the knowledge of Euclid's *Elements* and Apollonius' *Conics.*[27] The science of optics explains the reasons for errors in visual perception. Visual perception takes place through a cone formed by rays, in which the top is the point of vision and the base is the object seen. All objects appear larger if they are close to and smaller if they are distant from the observer. Furthermore, objects appear larger under water or behind transparent bodies.[28] Optics seek to explain these scientific phenomena by geometric means. Optics also presents an explanation of the differences in the perspective view of the moon at various latitudes. Knowledge of the phases of the moon and of the occurrence of eclipses is based on these conjectures.[29]

A great stimulus to optical investigation was provided in the first half of the eleventh century by Ibn Al-Haitham (Alhazen).[30] This Muslim mathematician was the first scholar to attempt to refute the optical doctrines to Euclid and Ptolemy. According to these doctrines, the eye received images of various objects by sending visual rays to certain points. In his book in optics, Al-Haitham proved that the process is actually the reverse and thus laid the foundations of modern optics. His formula was that it is not a ray that leaves the eye and meets the object that gives rise to vision, but rather that the form of the perceived object passes into the eye and is transmitted by the lens.[31]

Geometry was used extensively by Al-Haitham in his study of optics. His work on optics, which included one of the earliest scientific accounts of atmospheric refraction, contained a geometrical solution to the problem of finding the focal point of a concave mirror; that a ray from a given point must be incident in order to be reflected to another given point.[32] Al-Haitham also discovered some original geometrical theorems such as the theorem of the radical axis.[33]

The works of Ibn Al-Haitham became known in Europe during the twelfth and thirteenth centuries. Joseph ibn 'Aqnin referred to Ibn Al-Haitham's work in optics as being greater than those of Euclid and Ptolemy.[34] Al-Haitham's optics were made known to European mathematicians at about the same time by John Peckham, the Archbishop of Canterbury, in 1279, and by the Polish physicist, Witelo.[35]

Al-Haitham established the fundamental basis which eventually led to the discovery of magnifying lenses in Italy. Most of the medieval writers in the field of optics, including Roger Bacon, used his findings as their beginning. They particularly used *Opticae Thesaurus* (Compendium of Optics), Al-Haitham's book which was very important to Leonardo da Vinci and Johann Kepler.[36] During the seventeenth century Al-Haitham's work was very useful to the famous Kepler.[37] The writings of Al-Haitham are 'rooted in very sound mathematical knowledge, a knowledge that enabled him to propound . . . revolutionary doctrines on such subjects as the halo and the rainbow, eclipses and shadows, and on spherical and parabolic mirrors.'[38]

Prior to his death in Cairo, Al-Haitham issued a collection of problems similar to the *Data* of Euclid.[39] He is known to have written nearly two hundred works on mathematics, physics, astronomy, and medicine. He also wrote commentaries on Aristotle and the Roman physician, Galen. Although he made major contributions to the field of mathematics, it is especially in the realm of physics that he made his outstanding contributions. He was an accurate observer and experimenter, as well as a theoretician.[40] Howard Eves has observed:

> The name Al-Haitham . . . (965-1039), has been preserved in mathematics in connection with the so-called problem of Alhazen: To draw from two given points in the plane of a given circle lines which intersect on the circle and make equal angles with the circle at that point. The problem leads to a quartic equation which was solved in Greek fashion by an intersecting hyperbola and circle. Alhazen was born in Basra in South Iraq and was perhaps the greatest of the Muslim physicists. The above problem arose in connection with his optics, a treatise that later had great influence in Europe.[41]

The following is a partial list of Al-Haitham's works on geometry as appears in the *Thirteen Books of Euclid's Elements, Volume I:*

1. Commentary and abridgement of the *Elements*
2. Collection of the Elements of Geometry and Arithmetic, drawn from the treatises of Euclid and Apollonius
3. Collection of the Elements of the Calculus deduced from the principles laid down by Euclid in his *Elements*
4. Treatise on 'measure' after the manner of Euclid's *Elements*
5. Memoir on the solution of difficulties in Book I
6. Memoir for the solution of a doubt about Euclid, relative to Book V
7. Memoir on the solution of a doubt about the stereometric portion
8. Memoir on the solution of a doubt about Book XII
9. Memoir on the division of the two magnitudes mentioned

in Book X (Theorem of exhaustion)
10 Commentary on the definitions in the work of Euclid.[42]

Ibn Al-Haitham tried to prove Euclid's fifth postulate. The Greek's attempt to prove the postulate had become a 'fourth famous problem of geometry,' and several Muslim mathematicians continued the effort. Al-Haitham started his proof with a trirectangular quadrilateral (sometimes known as 'Lambert'd quadrangle' in recognition of Lambert's efforts in the eighteenth century). Ibn Al-Haitham thought that he had proved the fourth angle must always be a right angle. From this theorem on the quadrilateral, the fifth postulate is shown to follow. In his 'proof' he assumed that the locus of a point that remains equidistant from a given line is necessarily a line parallel to the given line, which is an assumption shown in modern times to be equivalent to Euclid's postulate.[43]

According to Hakim Mohammed Said, Presdent of Hamdard National Foundation in Karachi:

> In this year of grace, when man has first set foot on the moon and is reaching out to other stars, it is salutary to remember and acknowledge the great debt that modern mathematics and technology owe to the patient and exacting work of the early pioneers. This year we celebrate the 1,000 anniversary of one of the greatest of them, Abu Ali ibn Al-Hasan ibn Al-Haitham . . . Ibn Al-Haitham was a man of many parts, mathematician, astronomer, physicist, and physician. He had a 20th century mind in a 10th century setting and his contributions to knowledge were quite extraordinary.[44]

Thabit Ibn Qurra

Thabit ibn Qurra (836-911 AD) of Harran, Mesopotamia, is often regarded as the greatest Arab geometer.[45] He carried on the work of Al-Khwarizmi and translated into Arabic seven of the eight books of the conic sections of Apollonius.[46] He also translated certain works of Euclid, Archimedes, and Ptolemy which became standard texts.[47]

Archimedes' original work on the regular heptagon has been lost, but the Arabic translation by Thabit ibn Qurra proves the Greek manuscript still existed at the time of translation. Carl Schoy found the Arabian manuscript in Cairo, and revealed it to the Western public. It was translated into German in 1929.[48]

Ibn Qurra wrote several books on the subject of geometry. A partial list of his works includes: *On the Premises* (Axioms, Postulates, etc.) *of Euclid, On the Propositions of Euclid*, and a book on the propositions and questions which arise when two straight lines are cut by a third (the 'proof' of Euclid's famous postulate). He is also credited with *Introduction to the Book of Euclid*, which is a treatise on geometry.[49]

The starting point for all geometric studies among Muslims was Euclid's *Elements.*[50] Ibn Qurra developed new propositions and studied irrational numbers. He also estimated the distance to the sun and computed the length of the solar year.[51] He solved a special case of the cubic equation by the geometric method, to which Ibn Haitham had given particular attention in 1000 AD. This was the solution of cubic equations of the form $x^3 + a^3b = cx^2$ by finding the intersection of $x^2 = ay$ (a parabola) and $y(c - x) = ab$ (a hyperbola).[52]

Other Muslim Geometers

Al-Kindi

Al-Kindi, who made significant contributions in the field of arithmetic, also worked in the area of geometry. His most important contribution to scientific knowledge was his work on optics, dealing with the reflection of light, and his treatise on the concentric structure of the universe.[53] Using a geometrical model, Al-Kindi gave a 'proof' of the following:

1. The body of the universe is necessarily spherical
2. The earth will necessarily be spherical and (located) at the center of the universe
3. It is not possible that the surface of the water be non-spherical.[54]

Al-Kindi wrote many works on spherical geometry and its application to the universe. The following is a partial list of his works on spherics:

1. Manuscript on 'The body of the universe is necessarily spherical'
2. Manuscript on 'The simple elements and the outermost body are spherical in shape'
3. Manuscript on 'Spherics'
4. Manuscript on 'The construction of an azimuth on a sphere'
5. Manuscript on 'The surface on the water of the sea is spherical'
6. Manuscript on 'How to level a sphere'[55]
7. Manuscript on 'The form of a skeleton sphere representing the relative positions of the ecliptic and other celestial circles'.[56]

Al-Khwarizmi

Al-Khwarizmi's algebra also contained some geometrical ideas, according to Florian Cajori. He not only gave the theorem of the right triangle when the right triangle is isosceles, but also calculated the areas of the triangle, parallelogram, and circle. For π he used the approximation $3^1/_7$.[57] One chapter in Al-Khwarizmi's *Algebra* on mensuration dealt only with geometry and is called *Bab al-Misaha* (Chapter on measurement of areas).[58] If Al-Khwarizmi had really studied Greek mathematics, there would certainly have been some traces of the contents or terminology of Euclid's *Elements* in his geometry. There are none.[59] Euclid's *Elements* in their spirit and letter are entirely unknown to him.[60]

Al-Hajjaj ibn Yusuf

Al-Hajjaj ibn Yusuf, Muslim geometer, translated the *Elements of Euclid* for Harun al-Rashid (786-809 AD), renaming the work 'Haruni.' Al-Hajjaj revised his first translation for Al-Ma'mun (813-33 AD), the Caliph, and the revised work was known as Al-Ma'mun.[61]

The translation of the *Elements of Euclid* by Al-Hajjaj did not include Book X, which was later translated with Pappus' commentary by Sa'id ad-Dimishqi.[62]

Summary

The Muslims emphasized the study of geometry in their curriculum because it possessed practical applications in surveying, astronomy, and it aided the study of algebra and physics. Muslim geometry can be divided into constructional and arithmetical branches. When constructions were involved, the Muslims expressed the elements of geometrical figures in terms of one another, that is, by the methods of Greek geometry. Al-Khwarizmi was representative of this approach, with the solutions involving no arithmetical or algebraic technique. However, the numerical approach was more characteristic of Muslim geometry. According to Suter:

> In the application of arithmetic and algebra to geometry, and conversely in the solutions of algebraic problems by geometric means, the Muslims far surpassed the Greeks and Hindus.[63]

The work of Ibn Al-Haitham on optics was the outstanding Muslim work in the area of applied geometry. In his work, Al-Haitham challenged the doctrines of Euclid and Ptolemy. While using geometry most effectively, he also contributed to the development of the subject with his work on the radical axis. Thabit ibn Qurra's translation of Archimedes' work on the regular heptagon saved the manuscript from being lost forever. Ibn Qurra also contributed several original texts based on the work of Euclid, and he generalized the Pythagorean theorem.

Finally, as the signs of mathematical awakening of Europe appeared in the thirteenth century, the Greek classics were available for translation. As the Christian monks made contact with Muslim universities in Spain, opening the way to the Renaissance, Euclid's *Elements* were translated again, but this time from Arabic to Latin.

Notes

1. Marc Berge, *Risala Abi Hayyan Fi l-'Ulm: D' Abu Hayyan al-Tawhidi* (Paris, Extrait du Bulletin d'Etudes Orientales de L'Institut Francais De Damas Tome, XVIII, 1963-4), p. 289.

2. George Sarton, *Ancient Science and Modern Civilization* (Loncoln, Nebraska, University of Nebraska, 1954), p. 8.
3. University Library, Cambridge, England, Arabic MSS, 1075, fol. (00. 6.55).
4. Sir Thomas Arnold and Alfred Guillaume, *Legacy of Islam* (London, Oxford University Press, 1949), p. 380.
5. Howard Eves, *An Introduction to the History of Mathematics* (New York, Holt, Rinehart and Winston, 1969), p. 90.
6. Abd-ar-Rahman ibn Muhammad ibn Khaldun al-Hadrami, *The Muguaddemah's ibn Khaldun* (New York, Bollingen Foundation, 1958), pp. 130-1.
7. Shibli, *Recent Developments in the Teaching of Geometry* (York, Pennsylvania, The Maple Press Company, 1932), p. 16.
8. William David Reeve, *Mathematics for the Secondary School* (New York, Henry Holt and Company, 1954), p. 373.
9. Olinthus Gregory, *Mathematics for Practical Men* (Philadelphia, T. K. and P. G. Collins, 1838), p. 104.
10. James McMahon, *Elementary Plane Geometry* (New York, American Book Company, 1903), p. 1.
11. Edward Rutledge Robbins, *Plane Geometry* (New York, American Book Company, 1906), p. 11.
12. Charles S. Venable, *Elements of Geometry* (New York, University Publishing Company, 1875), p. 19.
13. H. A. Freebury, *A History of Mathematics: For Secondary Schools* (London, Cassell and Company, 1958), p. 32.
14. Hayward R. Alker, Jr., *Mathematics and Politics* (New York, The Macmillan Company, 1968), pp. 1-2.
15. Sir Thomas Heath, *A History of Greek Mathematics* (London, Oxford University Press, 1921), Vol. I, p. 355.
16. George Sarton, *Introduction to the History of Science* (Baltimore, The Williams and Wilkins Company, 1953), Vol. II, Part I, p. 9.
17. William David Reeve, 'The Teaching of Geometry,' *The National Council of Teachers of Mathematics*, Fifth Yearbook (New York, Teachers College, Columbia University, 1930), p. 1.
18. Howard Eves, *A Survey of Geometry* (Boston, Allyn and Bacon, 1963), Vol. I, p. 1.
19. M. A. Craig, *Al-Handasah Attahliliyah* (Cairo, Mutba'att Al-Ma'arif we Maktabatiha bi Masr, 1928), Vol. I, pp. 5-6.
20. David Eugene Smith, *History of Mathematics* (New York, Dover Publications, 1958), Vol. I, p. 81.
21. D. M. Y. Somerville, *The Elements of Non-Euclidean Geometry* (New York, Dover Publications, 1958), p. 1.
22. James B. Dodd, *Arithmetic* (New York, Pratt, Oakley and Company, 1857), p. 1.
23. A. Wilson Goodwing and Glen D. Vannatta, *Geometry* (Columbus, Ohio, Charles E. Merreill Books, 1961), p. 1.
24. Solomon Gandz, 'A Few Notes on Egyptian and Babylonian Mathematics,' *Studies and Essays in the History of Science and Learning* (Offered in Homage to George Sarton on the occasion of his sixtieth birthday) (New York, Henry Schumann, 1944), p. 460.
25. 'The Role of Mathematics in Civilization,' *The Place of Mathematics in Secondary Education*, Fifteenth Yearbook of the National Council

of Teachers of Mathematics (New York, Bureau of Publications of Teachers College, Columbia University, 1940), p. 3.
26. Benjamin Farrington, *Science in Antiquity* (London, Oxford University Press, 1947), p. 53.
27. A. C. Crombie, *Augustine to Galileo: The History of Science 400-1650* (London, The Falcon Press, 1952), p. 72.
28. Abd-ar-Rahman Ibn Muhammad Ibn Khaldun Al-Hadrami, *The Muquaddemah's Ibn Khaldun* (New York, Bollingen Foundation, 1958), Vol. III, p. 132.
29. Johannes Baamann, *Ibn al-Haitham's Abhandlung uber das Licht* (Leipzig, Halle Als, 1882), p. 37.
30. George Sarton, *Introduction to the History of Science* (Baltimore, The Williams and Wilkins Company, 1931), Vol. II, Part II, p. 761.
31. Sir Thomas Heath, *A History of Greek Mathematics* (London, Oxford University Press, 1921), Vol. II, pp. 293-5.
32. W. W. Rouse Ball, *A Short Account of the History of Mathematics* (New York, Dover Publications, 1960), pp. 161-2.
33. Mustafa Nazif Bik, *Al-Hasan ibn al-Haitham* (Buhuthuh wa-Kushufuh) (Cairo, Mutba'ah al-'timad bi-Masr, 1942), Vol. I, p. 9.
34. Sarton, *Introduction to the History of Science*, op. cit., Vol. II, Part II, p. 761.
35. Rene Taton, *History of Science* (New York, Basic Books, 1963), Vol. I, p. 482.
36. Heath, op. cit.
37. H. L. Kelly, 'History of Astronomy,' in Martin Davidson (ed.), *Astronomy for Every Man* (London, J. M. Dent and Sons, 1953), pp. 412-3.
38. Rom Landau, *Islam and the Arabs* (New York, The Macmillan Company, 1959), p. 185.
39. Ball, op. cit., p. 161.
40. Seyyed Hossein Nasr, *Science and Civilization in Islam* (Cambridge, Mass., Harvard University Press, 1968), p. 50.
41. Howard Eves, *An Introduction to the History of Mathematics* (New York, Holt, Rinehart and Winston, 1969), p. 194.
42. Sir Thomas L. Heath, *The Thirteen Books of Euclid's Elements* (New York, Dover Publications, 1956), Vol. I, pp. 88-9.
43. Kamal-Addin Abi al-Hasan al-Farisi, *Kitab Tangih al-Manazir* (Hyderabad, India, Bi Mutba 'att Majlis da' irat al-Ma'arif al-'Uthmaniyah, 1928), Vol. II, pp. 310-20.
44. Hakim Mohammed Said, 'Ibn al-Haitham Was a Bridge Between Ancient and Modern Sciences,' *Ibn al-Haitham* (Karachi, Pakistan, The Hamdard Academy Press, 1969), p. 29.
45. Carl Fink, *A Brief History of Mathematics* (Chicago, The Open Court Publishing Company, 1900), p. 320.
46. Arnold and Guillaume, op. cit., p. 387.
47. Francis J. Carmody, *The Astronomical Works of B. Kurra* (Berkeley, California, University of California Press, 1960), p. 15.
48. Robert W. Marks, *The Growth of Mathematics from Counting to Calculus* (New York, Bantam Books, 1964), p. 120.
49. Indian Office Library, London, England, Arabic MSS, 744, fol. 1^b-2^a.
50. Florian Cajori, *A History of Elementary Mathematics* (New York, The Macmillan Company, 1917), pp. 126-7.

51. Sydney N. Fisher, *The Middle East* (New York, Alfred A. Knopf, 1969), pp. 116-7.
52. David Eugene Smith, *History of Mathematics* (Boston, Ginn and Company, 1925), Vol. II, pp. 455-6.
53. Charles Singer, *A Short History of Scientific Ideas to 1900* (Glasgow, Clarendon Press, 1960), pp. 151-2.
54. Aydin Sayilik, 'Thabit Ibn Kurra's Generalization of the Pythagorean Theorem,' *Isis*, LI (March 1960), Part I, No. 163, 35-6.
55. George N. Atiyeh, *Al-Kindi: The Philosopher of the Arabs* (Karachi, Al-Karimi Press, 1966), pp. 166-8.
56. Sorbonne University, Paris, France, Arabic MSS, 2544 i bl. Gal. SI. 374.
57. Florian Cajori, *A History of Mathematics* (New York, The Macmillan Company, 1919), p. 104.
58. Indian Office Library, London, England, Arabic MSS, 750, fol. 41[b]-42[a].
59. Solomon Gandz, 'The Sources of Al-Khwarizmi's Algebra,' George Sarton (ed.), *Osiris*, I (1936), 264.
60. Solomon Gandz, *The Geometry of Muhammed ibn Musa al-Khwarizmi* (Berlin, Verlag von Julius Springer, 1932), p. 64.
61. George Sarton, *A History of Science* (Cambridge, Harvard University Press, 1959), Vol. II, p. 48.
62. De Lacy Evans O'Leary, *How Greek Science Passed to the Arabs* (London, Routledge and Kegan Paul, 1951), p. 158.
63. Taton, op. cit., p. 408.

7 CONCLUSIONS

The purpose of this work has been to present a brief history of Muslim contributions to mathematics. The study has concentrated upon the Golden age of Muslim culture – approximately 700 to 1300 AD. It was during this period of time that the Muslims came into possession of a spirit of discovery and scholarship which distinguished the era as the 'Muslim Renaissance' and led to setting the stage for the renaissance of learning in Europe, beginning about 1400.

I have presented a terse review of the historical, religious, and political settings for the era. Specific attention has been directed toward an introduction of the principal Muslim mathematicians and their chief contributions. To provide for an orderly presentation of mathematical contributions, I have directed attention to the topics of arithmetic, algebra, trigonometry, and geometry. Some effort has been devoted to a discussion of Muslim applied mathematics, particularly in the areas of physics and astronomy. No discussion has been presented with regard to medical and technological achievement of the Muslims.

The essential problem has been to describe these principal contributions to mathematics which are in evidence in Eastern and Western cultures at the present time. Selection of appropriate examples to illustrate the mathematical ideas developed by Muslim scholars of the era has presented another problem. But central to the narrative have been the ideas that (1) the Muslims preserved and extended the knowledge of their times and (2) modern civilization, as we know it today, would not exist without the Muslim Golden Age.

Review of Muslim Contribution to Mathematics

The study of science and mathematics was kept alive by the Muslims during a period when the Christian world was fighting desperately with barbarism. From the beginning of the eighth to the end of the thirteenth centuries, the Muslims acquired knowledge from the many varied sources available and disseminated this knowledge among the countries of the Mediterranean. Their favourable attitude toward the acquisition and development of knowledge attracted many Western scholars with a talent for science and a desire to study.

The investigations into the sciences was given impetus by the teachings of the Prophet Mohammed. His concern for knowledge was proclaimed in the Koran. This strong religious foundation provided a basis for the intense interest in education, and the desire to increase knowledge was shown by the Muslim scholars. The Caliphs who succeeded Mohammed continued to engender this positive attitude toward knowledge by establishing centers for learning, such as those in Cordova and Damascus.

Information from many cultures was collected by the Muslim scholars. The main patterns of mathematical thought are considered to be of Hellenistic origin, while astronomy and some mathematics came from Babylonian sources. The astronomy of the Greeks had been derived from the Babylonian and Egyptian cultures. Since the Babylonians had migrated from the Arabian peninsula to Mesopotamia, the astronomy which the Muslims restored to Europe was an achievement of their own cousins, an achievement borrowed by Greeks many centuries earlier. However, the Muslims not only collected knowledge from the Greeks and Eastern sources, but they provided many original contributions. As George Sarton observes:

> To return to the Renaissance, it was, among other things, a revolt against medieval concepts and methods. Of course every generation reacts against the former one; every historical period is a revolt against its predecessor. Yet in this case the revolt was sharper than usual. It is not sufficiently realized that the Renaissance was not simply a revolt against scholasticism; it was also directed against Arabic influences.
>
> The anti-Arabic drive was already in full swing in Petrarch's time. Such a revolt and struggle for independence were symptoms of growing strength. The revolt was successful, but not complete; there are still many Arabic elements in our language and culture.[1]

Great service to the progress of civilization was rendered by the Muslim scholars, not only by their writing of textbooks in the areas of arithmetic, algebra, and trigonometry, but by their scholarly development of these subjects of mathematics. Muslim mathematicians proved the value of numbers to civilization by relating arithmetic to everyday life. They produced the system of Arabian numerals, which replaced the awkward system of Roman numerals, and they made several discoveries concerning 'amicable' numbers. The invention of the proof by 'casting out the nine' is attributed to them, as well as the rule of double false position (*regula duorum falsorum*). This practice is still in use in modern Saudi Arabia.

During their period of enlightenment, Muslims made significant contributions to the development of algebra and our present number system. The Muslim West introduced a symbolic algebra as evidenced by the work of the Andalusian writer, Alkalsadi. In the title of his book, *Raising the Veil of Science of Gubar,* the term *gubar* was used to present *written* as distinguished from *mental* arithmetic.

The science of light was studied by the Muslims, who appear to have been the first to make lenses. This later art was found to be very useful by Galileo. The Muslims built many astronomical instruments of the types that are still in use today. However, they added little to the geometry formulated by the Greeks, but did lay the foundations for analytic geometry and were the founders of plane and spherical trigonometry.

Algebra is a Muslim science. Diophantus of Alexandria had discussed the solution of first- and second-degree equations, but he lacked an easy system of numerals on which to build his solutions. The Muslims used the ideas of Diophantus and of the Indians together with their own system of numerals to develop algebra and introduce its name. They initiated the use of algebraic symbols taken from the letters of their alphabet.

The Muslims also discovered the connection between algebra and geometry, using algebraic methods to solve geometric problems. Thus, they laid the foundations for analytical geometry. The Muslims were the first to use the method of successive approximations in solving problems, and thus they provided a basis for numerical methods.

Trigonometry was developed to a great extent by the Muslims. They systematized the contributions of the Alexandrians and Indians, added considerable contributions of their own in plane and spherical trigonometry, and proceeded to establish the subject as a science independent of astronomy. They also made different tables leading to the discovery of the law of logarithms six hundred years before John Napier was credited with its invention.

In astronomy itself the Muslims improved the Greek astrolabe and invented many accurate instruments for studying the stars and measuring angular distances between celestial bodies. As a result of their studies, the Muslims established the fact that the earth is a sphere floating in space. They performed a highly complicated geodetic operation to calculate the length of a terrestrial degree and used the result to determine the circumference and diameter of the earth.

The Muslims produced a number of great mathematicians, primarily during the ninth and tenth centuries. Some of the most famous are Thabit ibn Qurra, Al-Battani, Al-Biruni, Al-Kindi, and Al-Khwarizmi, the 'father of algebra.'

In estimating the importance of the Muslim contribution to mathematics, one must understand that a systematic method of analysis and an easy means of numerical calculation are prerequisites for any culture. None of the modern sciences could have flourished without the Arabian system of calculation and the science of algebra. The use of Arabic numerals, particularly the zero, made possible the solution of long, involved equations. Thus, to the Muslim mathematicians goes the credit for making possible the vast development of modern science.

The Decline of Muslim Influence

The achievements of Muslim culture at a time when most of Europe endured intellectual darkness, religious intolerance, and crudeness of customs must seem strange in the twentieth century. The present Muslim world remains far behind the West, although in religion, race, geographical surroundings, language, and historical background, the Muslims of today do not differ much from their ancestors in the days of the Abassids. The Muslim Empire and its civilization remained the dominant force in the Western world for half a millennium. Even after the thirteenth century the civilization produced great achievements for another two hundred years.

The downfall of the Empire might have been slowed down had

it not been for the growing political disunity created by the personal ambitions and jealousies that were characteristic of medieval history, both Muslim and Christian. The Mongol invasion was the force which struck the fatal blow to Muslim civilization. Cities, monuments, and orchards were razed to the ground, and people were massacred by the thousands. In the aftermath of the invasion, the Muslims were unable to recapture their former power or regain their former accomplishments.

Near the end of the fifteenth century, the Muslims came under the influence of the Ottoman Turks. Although the Ottomans were converts to Islam, they were different from Arab Muslims by race, historical background, temperament, and language. It was they who became the rulers of the Middle East and most of North Africa. The Arabian Muslims, therefore, held the status of a colonized people. The Turks considered the Arab Muslims as inferior, and the whole population was excluded from all activities that contribute to a genuine civilization. The Ottomans cut themselves off from cultural interchanges with the Western world and did not participate in Europe's rapid cultural advance.

The only Western ideas that were permitted to enter the cultural barriers erected about the Ottoman Empire were those of warfare. During the decisive days of the Renaissance, of the eighteenth century Enlightenment, and of the Industrial Revolution, the Arab Muslims were made to perform menial tasks as wood cutters and carriers of water. The political and economic conditions within the Ottoman Empire enabled only a small number of Arab Muslims to derive benefit from cultural centers; therefore, apathy and discouragement were widespread. In the nineteenth century certain elements of Western culture succeeded in pushing through the barriers of the Ottoman seclusion.

In the nineteenth century, Egypt became almost independent from Constantinople, but this freedom did not last long. By 1880 Great Britain had begun an era of Western colonialism in the Arab Muslim countries. The majority of Arab Muslims knew no real independence until after the Second World War, and even then some still lived under foreign rule. It was only in 1956 that Morocco and Tunisia became independent. Even after that date French military and economic control still existed.

People who have lived under foreign domination for nearly four hundred years can hardly be expected to achieve the same degree of progress it has taken the Western world centuries to consolidate. The newly-independent Arab Muslim States also have had to live with complications brought about by unwelcomed foreign leadership. The establishment of a Zionist State in the center of an Arab Muslim world has created endless tensions and, ultimately, wars. Most of the energies of the Arab Muslims have been released into political activity rather than into social and cultural pursuits.

The Arabic language is considered a strong bond which holds the Arabs together. Arabic is the mother tongue of the people living in countries from Morocco to Iraq. Although the local dialects vary, there is but one classical form of the language, and the dialects are easily understood by the inhabitants of the various regions. A deep attachment to the Arabic language is characteristic of all Muslim Arabs; the beauty and rhythm of the language stir them deeply.

A New Challenge for Present-Day Muslim Thought

René Maheu, the Director General of the United National Educational, Scientific, and Cultural Organization, in speaking of the tremendous development in education in the Muslim countries, has said that during the ten years between 1950 and 1960, education has developed rapidly at all levels in the Muslim countries. The number of children enrolled in primary schools has almost doubled. The enrollment of children in secondary establishments is nearly three times higher than in 1956 while the number of students in the major universities has doubled.[2]

The late President of the United States of America, Dwight D. Eisenhower, in an address before the United Nations General Assembly, stated:

> As I look into the future I see the emergence of modern Muslim States that would bring to this century contributions surpassing those we cannot forget from the past. We remember that Western arithmetic and algebra owe much to Muslim mathematicians and that much of the foundation of the world's medical science and astronomy was laid by Muslim scholars. Above all, we remember that three of the world's great religions were born in the Near East.[3]

So far as twentieth century civilization is concerned, the Muslim world is young. There is no reason to assume that the Muslims have lost the qualities which enabled their ancestors to evolve their rich civilization. The Muslims must concentrate on raising their standards of education and deepening their sense of social justice. These basic tasks will leave them less energy for cultural pursuits, although such an undertaking is a contribution to culture itself.

The religion of Islam is an incentive to progress. In the past few hundred years, Islam has been subjected to enlightened influences and to reform movements that have recognized the necessity for adapting the old teaching to the present era. There are many Muslim thinkers, planners, and writers whose advocacy of progress with Western ideas is reconcilable with their allegiance to Islam. Such a view suggests the possibility of a change for Muslim cultural perspectives and a return to scholarly pursuits in education and applied research.

Recorded history takes note that a score of great civilizations have risen and fallen over the past five millenia. None of these civilizations, which rose to pinnacles of greatness, has ever again achieved its former glory. Muslim civilization can become an exception to this by a re-examination of its past contributions to Eastern and Western cultures and by a willingness to once again 'span the gap' for the diverse people of the world.

Recommendations for Further Study

The present work has been intentionally brief. It is not possible to do justice to the Muslim contributions to mathematics in such a condensed treatment. Not only is the subject a vast one, but proper research into the original manuscripts that are presently available presents a serious problem for the researcher in terms of time, effort, and the location of materials. Translation does not present a real problem for the researchers since most of the available manuscripts are now written in Arabic.

Perhaps the best sources for these translations, as well as for the original manuscripts, are the libraries of the Indian Office, the British Museum, and the University of Cairo. At these institutions one may find extensive Muslim works which date back to the seventh century. Although few of these manuscripts have been translated into English, one may still have access to the original manuscripts.

There are many suggestions for further study in these Muslim works. Researchers will find a rich lore of ideas, far beyond those suggested in the present work, which will provide a considerable challenge in terms of such things as the advanced nature of mathematical thinking during the Muslim Golden Age. One very important area of the author's research, number theory, is suggested as a fruitful field for further investigation. The present investigation has touched upon a few of these ideas in Chapter 2. However, there is a wealth of material in the field of number theory which could provide the basis of a study of this branch alone of Muslim contributions to mathematics.

Another field worth investigation is the successful attempt on the part of Muslim mathematicians to show a definite relationship between the fields of geometry and algebra. This effort was a prelude to analytic geometry, and a brief discussion of this was given in Chapter 5.

The close relationships that exist between Muslim trigonometry and astronomy, together with their effects on later works by European scholars, provide for fruitful excursions into some very fundamental ideas which have also influenced the rise of modern science and technology. For example, modern astronomy owes much to the work of the Muslim scholars in the field of trigonometry, as well as the Muslim study of optics and the construction of lenses.

Still further areas in which further research can be useful are those of the Muslim application of mathematics to physics, medicine, chemistry, pharmacology, and agriculture. It must be recognized that, in the main, the Muslims were practical people with an interest in the application of ideas to the improvement of their culture. Thus, it would seem reasonable to suggest that such a study would provide insight into the very nature of how theory may become practice, and how application may promote generalization of ideas to more theoretical investigations. To say the least, the interaction of theory versus application has always played an important role in the development and decline of civilizations. Such a study as this could suggest ways for our present culture to come to grips with how best to establish a balance between these apparently complementary aspects of man's endeavors.

Notes

1. George Sarton, *Six Wings: Men of Science in the Renaissance* (Bloomington, Indiana University Press, 1957), pp. 3-4.
2. Rene Maheu, 'The Development of Education in the Arab Countries,' *UNESCO Chronicle*, VI, No. 4 (April 1960), 137.
3. President Dwight D. Eisenhower, 'Address to the United Nations General Assembly,' *Arab World*, V, Nos. 1-2 (January-February, 1959), 2.

BIBLIOGRAPHY

A. Books

Addahan, Sa'id Nasir, *Al-Kur'an wa Al-'Ulum* (Karbala, Mutba'at Anna'man, 1965)

Agil, Nabih, *Tarikh Al-Arab Al-Gadim* (Damascus, Damascus University Library, 1968)

Al-Bakhari, Abu 'Abdullah Muhammad B. Isma'il, Al-Imam, *At-To-Rikhu I-Kabir: A Dictionary of the Biography of Traditionists* (Hyderabad, India, Osmania Oriental Publications Bureau, 1942), Vol. I, Part I

Al-Basiri, Muhammed Mahdi, *Al-Muwashah fi Al-Andalus wa-fi Asharg* (Baghdad, Mutba'at Al-Ma'arif, 1948)

Al-Biruni, Abul Rihan Mohammed ibn Ahmen, *Tahdid al-Amakin* (Beirut, The American University Press, 1966) (trans. Jamil Ali

Al-Farisi, Kamal-Addin Abi al-Hasan, *Kitab Tangih al-Manazir* (Hyderabad, India, Bi Mutba'att Majlis Da'irat al-Ma'arif al-'Uthmaniyah, 1928), Vol. II

Al-Hilali, Taki Ed Din, *Die Einleitung Zu al-Birunis Steinbuch* (Leipzig, Otto Harrassowitz, 1941)

Al-Hurani, Abi 'Abdullah Muhammed bin Sinan bin Jabir, *Kitab Azzij Assabi'* (Rome, Tabi'a bi Madinat Rumiyan Al-Uzma, 1899)

Al-Ka'ak, 'Uthman, *Al-Haddarah Al-'Arabiyah fi Hawdd Al-Bahr Al-'Abyad* (Cairo, Jami'att Addiwal Al-'Arabiyah, 1965)

Alker, Hayward R., Jr., *Mathematics and Politics* (New York, The Macmillan Company, 1968)

Al-Ma'ruf bi 'ibn Al-'Athir, 'Abi Al-Hasan Ali ibn Muhammed, *Al-Kamil fi Attarikh* (Cairo, 'Idarat Attiba'at Al-Muniriyah bi Masr, 1929), Vol. II

Al-Nadim, Ibn, *Al-Fahrasat Li Ibn Al-Nadim* (Cairo, Al-Haj Mustafa Muhammed, 1800)

Al-Shuyashi, Muhammed Mufud, *Al-Arab wa-al-Hadaratt al-'urubiyah* (Cairo, Mutabi' Dar al-Galam, 1971)

Al-Tawil, Tawfiq, *Al-'Arab wa Al-'ilm* (Cairo, Maktabat Al-Nahdah

Al-Misriyan, 1968)

'Anan, Muhammed Abdullah, *Mawakif Hasimah fi Tarikh al-Islam* (Cairo, Mu'asasah al-Khaniji, 1962)

Arabian American Oil Company, *Aramco Handbook: Oil and the Middle East* (Netherlands, Joh. Enschede en Zonen-Haarlem, 1968)

Arnold, Thomas, Sir, and Guillaume, Alfred, *The Legacy of Islam* (London, Lowe and Brydone, 1949)

Ashati, Ahmed Shukut, *Maqmu't 'Abhath fi al-Hadarat al-Arabiat al-'Islamyyah* (Dimashg, Mutba't Gami'at Dimashg, 1963)

Assufi, Khalid, *Tarikh Al-'Arab fi 'Asbanya* (Damascus, Al-Mutba'ah Atta'awiniyah, 1959)

Assufi, Khalid, *Tarikh Al-'Arab fi Asbanya* (Nihayah Al-Khilafat Al-'Umawiyah fi Al-'Ndalus) (Halabb, Maktabat Dar Ashsharg, 1963)

Atiyah, Edward S., *The Arabs: The Origins, Present Conditions, and Prospects of the Arab World* (Edinburgh, R. and R. Clark, 1958)

Atiyah, George N., *Al-Kindi: The Philosopher of the Arabs* (Karachi, Al-Karimi Press, 1966)

Baamann, Johannes, *Ibn al-Haitham's Abhandlung Uber das Licht* (Leipzig, Halle Als., 1882)

Ball, W. W. Rouse, *A Primer of the History of Mathematics* (London, Macmillan and Company, 1927)

_____ *A Short Account of the History of Mathematics* (New York, Dover Publications, 1960)

Banks, J. Houston, *Elements of Mathematics* (Boston, Allyn and Bacon, 1969)

Baumgart, John R., 'History of Algebra,' *Historical Topics for the Mathematics Classroom*, Thirty-First Yearbook of the National Council of Teachers of Mathematics (Washington, D.C., National Council of Teachers of Mathematics, 1969)

Bell, E. T., *Men of Mathematics* (New York, Simon and Schuster, 1937)

_____ *The Development of Mathematics* (New York, McGraw-Hill Book Company, 1940)

Berge, Marc, *Risala Abi Hayyan Fi L-'Ulm: D'Abu Hayyan al-Tawhidi* (Paris, Extrait du Bulletin d'Etudes Orientales de L'Institut Francais De Damas Tome, XVIII, 1963-4)

Bernal, John Desmond, *Science in History* (London, C. A. Watts and

Company, 1957)

Bik, Mustafa Nazif, *Al-Hasan ibn al-Haitham: Buhuthuh wa-Kushufuh* (Cairo, Mutba'ah Al-'timad bi-Masr, 1942), Vol. I

Boas, Marie, *History of Science* (Washington, D.C., The American Historical Association, 1958)

Bochner, Solomon, *The Role of Mathematics in the Rise of Science* (Princeton, New Jersey, Princeton University Press, 1966)

Boyer, Carl B., *A History of Mathematics* (New York, John Wiley and Sons, 1968)

Boyer, Lee Emerson, *Mathematics: A Historical Development* (New York, Henry Holt and Company, 1949)

Briffault, Robert, *Rational Evolution* (New York, The Macmillan Company, 1930)

_____ *The Making of Humanity* (New York, The Macmillan Company, 1930)

Brockelmann, Carl, *History of the Islamic Peoples* (Cornwall, New York, The Cornwall Press, 1947)

Bush, George, Rev., *Life of Mohammed: Founder of the Religion of Islam and the Empire* (Niagara, Henry Chapman, 1831)

Byng, Edward J., *The World of the Arabs* (Boston, Little, Borwn and Company, 1944)

Cajori, Florian, *A History of Elementary Mathematics* (New York, The Macmillan Company, 1917)

_____ *A History of Mathematical Notations* (La Salle, Illinois, The Open Court Publishing Company, 1928)

_____ *A History of Mathematics* (London, Macmillan and Company, 1925)

Cantor, Norman F., *Medieval History: The Life and Death of a Civilization* (New York, The Macmillan Company, 1963)

Carmichael, Joel, *The Shaping of the Arabs: A Study in Ethnic Identity* (London, Collier Macmillan, 1967)

Conant, Levi Leonard, *The Number Concept* (New York, Macmillan and Company, 1923)

Craig, M. A., *Al-Handasah Attahliliyah* (Cairo, Matba'att Al-Ma'arif wa Maktabatiha bi Masr, 1928), Vol. I

Crichton, Andrew, *The History of Arabia: Ancient and Modern* (New York, Harper and Brothers, 1937), Vol. I

Crombie, A. C., *Augustine to Galileo: The History of Science 400-1650* (London, The Falcon Press, 1952)

Dampier, Sir William Cecil, *History of Science* (New York, The Macmillan Company, 1942)

—— *A Shorter History of Science* (New York, The Macmillan Company, 1945)

Dantzig, Robias, *Number, The Language of Science* (Garden City, New York, Doubleday and Company, 1956)

Datta, Bibhutibhusan, and Singh, Avadhesh Narayan, *History of Hindu Mathematics* (Lahore, Motilal Banarsi Das, 1935), Part I

Davis, William Stearns, *A Short History of the Near East: From the Founding of Constantinople* (New York, The Macmillan Company, 1922)

De Vaux, Carra, 'Astronomy and Mathematics,' *The Legacy of Islam* (London, Oxford University Press, 1931)

Dickson, Leonard Eugene, *History of Numbers* (Washington, D.C., Press of Gibson Brothers, 1919), Vol. I

Dood, James B., *Arithmetic* (New York, Pratt, Oakley and Company, 1857)

El-Azzawi, Abbas, *History of Astronomy in Iraq* (Baghdad, Iraq Academy Press, 1959)

Eves, Howard, *A Survey of Geometry* (Boston, Allyn and Bacon, 1963), Vol. I

—— *An Introduction to the Foundations and Fundamental Concepts of Mathematics* (New York, Rinehart and Company, 1958)

—— *An Introduction to the History of Mathematics* (New York, Holt, Rinehart and Winston, 1969)

Farrington, Benjamin, *Science in Antiquity* (London, Oxford University Press, 1947)

Fehr, Howard Franklin, *A Study of the Number Concept of Secondary School Mathematics* (Ann Arbor, Michigan, Edwards Brothers, 1945)

Feuer, Lewis Samuel, *The Scientific Intellectual: The Psychological and Sociological Origins of Modern Science* (New York, Basic Books, 1963)

Fine, Henry B., *Number-System of Algebra* (New York, D. C. Heath and Company, 1890)

Fink, Karl, *A Brief History of Mathematics* (Chicago, The Open Court Publishing Company, 1900)

Fisher, Sydney N., *The Middle East* (New York, Alfred A. Knopf,

1969)

Freebury, H. A., *A History of Mathematics* (New York, The Macmillan Company, 1961)

_____ *A History of Mathematics: For Secondary Schools* (London, Cassell and Company, 1958)

Gabrieli, Francesco, *The Arabs: A Compact History* (New York, Hawthorn Books, 1963)

Gallagher, Charles F., *A Note on the Arab World* (New York, American Universiry Field Staff, 1961)

Gamir, Yuhana, *Falasifat Al-'Arab* (Beirut, Al-Maktabat Ash Shargiyah, 1957)

Gandz, Solomon, 'A Few Notes on Egyptian and Babylonian Mathematics,' *Studies and Essays in the History of Science and Learning: Offered in Homage to George Sarton on the Occasion of his Sixtieth Birthday* (New York, Henry Schumann, 1944)

_____ *The Geometry of Muhammad ibn Musa al-Khwarizmi* (Berlin, Verlag Von Julius Springer, 1932)

Gibb, Hamilton Alexander Rosskeen, *An Interpretation of Islamic History* (Lahore, M. Ashraf Darr for Orientatia Publishers, 1957)

_____ *Studies on the Civilization of Islam* (Boston, Beacon Books on World Affairs, 1962)

_____ and Bowen, Harold, *Islamic Society and the West* (London, Oxford University Press, 1950), Vol. I, Part I

_____ and Bowen, Harold, *Islamic Society and the West* (London, Oxford University Press, 1957), Vol. I, Part II

Glubb, John Bagot, *The Course of Empire: The Arabs and Their Successors* (London, St. Paul's House, 1965)

_____ *The Life and Times of Mohammed* (New York, Stein and Day Publishers, 1970)

Glubb, Sir John Bagot, *The Great Arab Conquests* (Englewood Cliffs, New Jersey, Prentice-Hall, 1964)

Goodwin, Wilson, and Vanatta, Glen D., *Geometry* (Columbus, Ohio, Charles E. Merrill Books, 1964)

Gregory, Olinthus, *Mathematics for Practical Men* (Philadelpha, T. K. and P. G. Collins, 1838)

Guillaume, Alfred, *Islam* (Edinburgh, Britain, R. and R. Clark, 1954)

Hamady, Sania, *Temperament and Character of the Arabs* (New York, Twayne Publishers, 1960)

Harvey-Gibson, R. J., *Two Thousand Years of Science* (New York, The Millzn Company, 1929)

Hassan, Hassan Ibrahim, *Islam: A Religious, Political, Social, and Economic Study* (Baghdad, Iraq, The Times Printing and Publishing Company, 1967)

Hassler, Josper O., and Smith, Rolland R., *The Teaching of Secondary Mathematics* (New York, The Macmillan Company, 1935)

Heath, Sir Thomas, *A History of Greek Mathematics* (London, Oxford University Press, 1921), Vol. I

——— *A History of Greek Mathematics* (London, Oxford University Press, 1921), Vol. II

——— *The Thirteen Books of Euclid's Elements* (New York, Dover Publications, 1956), Vol. I

Hell, Joseph, *The Arab Civilization* (London, W. Heffer and Sons, 1943)

Historical Section of the Foreign Office, *Mohammedan History: The Rise of Islam and the Pan Islamic Movement* (London, H. M. Stationery Office, 1920), Vol. X

Hitti, Philip K., *History of the Arabs: From the Earliest Time to the Present* (London, Macmillan and Company, 1964)

——— *Makers of Arab History* (New York, Harper and Row Publisher, 1968)

——— *The Arabs: A Short History* (London, Macmillan and Company, 1948)

——— *The Near East in History – A 5000-Year Story* (New York, D. Van Nostrand Company, 1960)

Hocker, Sidney G., Barnes, Wilfrid W., and Long, Calvin T., *Fundamental Concepts of Arithmetic* (Englewood Cliffs, New Jersey, Prentice-Hall, 1963)

Hogben, Lancelot, *Mathematics for the Millions* (New York, W. W. Norton and Company, 1946)

Holt, R. M., Lambton, Ann K. S., and Lewis, Bernard (eds.), *History of Islam* (London, Cambridge University Press, 1970), Vol. I

Hooper, Alfred, *Makers of Mathematics* (New York, Random House, 1948)

——— *The River Mathematics* (New York, Henry Holt and Company, 1945)

Hottinger, Arnold, *The Arabs: Their History, Culture, and Place in the Modern World* (Berkeley and Los Angeles, University of

California Press, 1963)

Howe, George, *Mathematics for the Practical Man* (New York, D. Van Nostrand Company, 1957)

Hutton, Charles, *A Course of Mathematics* (Glasgow, Richard Griffin and Company, 1833)

Ibn Khaldun al-Hadrami, Abd-ar-Rahman ibn Muhamad, *The Muquaddemah's ibn Khaldun* (New York, Bollingen Foundation, 1958), Vol. I

Jurji, Edward J., *The Arab Heritage* (New Jersey, Princeton University Press at Princeton, 1944)

Karpinski, Louis Charles (trans.), *Robert of Chester's Latin Translation of Algebra of al-Khwarizmi* (London, Macmillan and Company, 1915)

Karpinski, Louis Charles, and Winter, John Garrett, *Contributions to the History of Science* (Ann Arbor, University of Michigan, 1930)

Kasir, Daoud S., *The Algebra of Omar Khayyam* (New York, J. J. Little and Ives Company, 1931)

Kasner, Edward, and Newman, James, *Mathematics and the Imagination* (New York, American Book-Stanford Press, 1945)

Kelly, H. L., *Astronomy for Every Man* (London, J. M. Dent and Sons, 1953)

Kennedy, Edward S., 'The History of Trigonometry,' *Historical Topics for the Mathematics Classroom*, Thirty-First Yearbook, National Council of Teachers of Mathematics (Washington, D.C., National Council of Teachers of Mathematics, 1969)

Khadduri, Majid, *The Law of War and Peace in Islam* (London, Luzac and Company, 1941)

Khan, Mohammad Abdur-Rahman, *A Brief Survey of Moslem Contribution to Science and Culture* (Lahore, Sh. Umar Daraz at the Imperial Printing Works, 1946)

Kline, Morris, *Mathematics and the Physical World* (New York, Thomas Y. Crowell Company, 1959)

Kokomoor, Franklin W., *Mathematics in Human Affairs* (New York, Prentice-Hall, 1946)

Kramer, Edna E., *The Main Stream of Mathematics* (New York, Oxford University Press, 1951)

——— *The Nature and Growth of Modern Mathematics* (New York, Hawthorn Books, 1970)

Landau, Rom, *Arab Contribution to Civilization* (San Francisco, The American Academy of Asian Studies, 1958)

_____ *Islam and the Arabs* (New York, The Macmillan Company, 1959)

_____ *The Arab Heritage of Western Civilization* (New York, Arab Information Center, 1962)

Lenczowski, George, *The Middle East in World Affairs* (Ithaca, New York, Cornell University Press, 1956)

Levey, Martin, *The Algebra of Abu Ramil* (The University of Wisconsin Press, 1966)

Lewis, Bernard, *The Arabs in History* (New York, Hutchinson's University Library, 1950)

Lindquist, Theodore, *Modern Arithmetic Methods and Problems* (Chicago, Scott, Foresman, and Company, 1917)

Linton, Ralph, *The Tree of Culture* (New York, Alfred A. Knopf, 1955)

Logsdon, Mayme I., *A Mathematician Explains* (Chicago, The University of Chicago Press, 1935)

Mahdi, Muhsin, *Ibn Khaldun's Philosophy of History: A Study in the Philosophic Foundation of the Science of Culture* (Chicago, The University of Chicago Press, 1964)

Marks, Robert W., *The Growth of Mathematics from Counting to Calculus* (New York, Bantam Books, 1964)

Marmery, J. Willin, *Progress of Science* (London, Chapman and Hall, 1895)

Marriman, Gaylord M., *To Discover Mathematics* (New York, John Wiley and Sons, 1942)

Mashrafah, Ali Mustafa, and Ahmad, Mohammed Musa (eds.), *Kitab al-Jabr wa al-Muqabalah li Mohammed ibn Musa al-Khwarizmi* (Cairo, Dar al-Katib al Arabi Littiba wa al-Nashr, 1968)

Mason, Stephen F., *A History of the Sciences* (New York, Collier Books, 1962)

Mazhar, Isma'il, *Tarikh al-Fikr al-'Arabi: fi Nushu'ih wa Tatawirih bi Ttarjamah wa Annagil 'an al-Hadarah al-Yunaniyah* (Cairo, Dar al-'Uswr li Ttab' wa Annashur bi Masr, 1928)

McMahon, James, *Elementary Plane Geometry* (New York, American Book Company, 1903)

McKenzie, A. E. E., *The Major Achievements of Science* (London,

Cambridge University Press, 1960)

Meakin, Budgett, *Moorish Empire: A Historical Epitome* (London, Swan Sonnenschein and Company, 1899)

Merrick, Donald, *Mathematics for Liberal Arts Students* (Boston, Princle, Weber and Schmidt, 1970)

Midonick, Henrietta O. (ed.), *The Treasure of Mathematics* (New York, Philosophical Library, 1965)

Miller, George A., *Historical Introduction to Mathematical Literature* (New York, The Macmillan Company, 1916)

Morgan, Kenneth W., *Islam – The Straight Path* (New York, The Ronald Press Company, 1958)

Muir, Jane, *Of Men and Number: The Story of the Great Mathematicians* (New York, Dodd, Mead and Company, 1961)

Nasr, Seyyed Hossein, *Science and Civilization in Islam* (Cambridge, Mass., Harvard University Press, 1968)

Neugebauer, O., *The Exact Sciences in Antiquity* (Providence, Rhode Island, Brown University Press, 1957)

O'Leary, DeLacy Evans, *How Greek Science Passed to the Arabs* (London, Routledge and Kegan Paul, 1951)

Ore, Oystein, *Number Theory and Its History* (New York, McGraw-Hill Book Company, 1948)

Pledge, H. T., *Science since 1500: A Short History of Mathematics, Physics, Chemistry, and Biology* (New York, Philosophical Library, 1947)

Poole, Lane, *The Story of the Moors in Spain* (London, G. P. Putnam's Sons, 1902)

Rahman, Fazlur, *Islam* (New York, Holt, Rinehart, and Winston, 1966)

Rangrut, *The Ideologies in Conflict* (Karachi, Central Printing Press, 1964)

Reeve, William David, *Mathematics for the Secondary School* (New York, Henry Holt and Company, 1954)

_____ 'The Teaching of Geometry,' *The National Council of Teachers of Mathematics*, Fifth Yearbook (New York, Teachers College, Columbia University, 1930)

Robbins, Edward Rutledge, *Plane Geometry* (New York, American Book Company, 1906)

_____ 'The Role of Mathematics in Civilization,' *The Place of Mathematics in Secondary Education*, Fifteenth Yearbook of the

National Council of Teachers of Mathematics (New York, Bureau of Publication of Teachers College, Columbia University, 1940)

Rosenthal, Franz (trans.), *The Muqaddimah ibn Khaldun: Autobiography* (New York, Bollingen Foundation, 1958), Vol. III

Sachau, C. Edward, *Chronologic Orientalischer Volker, Van al-Beruni* (Leipzig, In Commission Bei F. A. Brockhaus, 1878)

Saffauri, Mohammed, and Ifram, Adnan (trans.), *Al-Biruni on Transits* (Beirut, American University of Beirut Press, 1959)

Safwat, Ahmid Zaki, *Jamharat Rasa'il al-'Arab fi 'usur al-'Arabiyah Azzahirah* (Cairo, Sharikat Maktabat wa Mutba'at Mustafa al-Babi al-Halabi wa-'wladuh bi Masr, 1937)

——— *Jamharat Rasa'il al-'Arab fi 'usur al-Arabiyah Azzahirah* (Cairo, Sharikat Maktabat wa-Mutba'at Mustafa al-Babi al-Halabi wa-'wladuh bi Masr, 1937)

Said, Hakim Mohammad, *Ibn al-Haitham Was a Bridge Between Ancient and Modern Science* (Karachi, Pakistan, The Hamdard Academy Press, 1969)

Sanford, Vera, *A Short History of Mathematics* (New York, Houghton Mifflin Company, 1930)

Sarton, George, *A Guide to the History of Science* (Waltham, Mass., The Chronicle Botanica Company, 1952)

——— *A History of Science* (Cambridge, Mass., Harvard University Press, 1952), Vol. I

——— *Ancient Science and Modern Civilization* (Lincoln, Nebraska, University of Nebraska Press, 1954)

——— *Introduction to the History of Science: From Homer to Omar Khayyam* (Baltimore, The Williams and Willkins Company, 1953) Vol. I

——— *Introduction to the History of Science: From Rabbi Ben Ezra to Roger Bacon* (Baltimore, The Williams and Wilkins Company, 1953), Vol. II, Part I

——— *Introduction to the History of Science* (Baltimore, The Williams and Wilkins Company, 1931), Vol. II, Part II

——— *Introduction to the History of Science* (Baltimore, The Williams and Wilkins Company, 1948), Vol. III, Part II

——— *Six Wings: Men of Science in the Renaissance* (Bloomington, Indiana University Press, 1957)

——— *The Appreciation of Ancient and Medieval Science During the Renaissance: 1450-1600* (Philadelphia, University of Pennsylvania Press, 1955)

Sarton, George, *The Incubation of Western Culture in the Middle East* (Washington, D.C., The Library of Congress, 1951)

_____ *The Life of Science: Essays in the History of Civilization* (New York, Henry Schumann, 1948)

Saunders, John Joseph, *A History of Medieval Islam* (London, Routledge and Kegan Paul, 1966)

Sayed, Ali Amir, *Mukhtasar Tarikh Al-'Arab* (Beirut, Dar Al-Galam Limalayin, 1961)

Sayili, Aydin, *'Abd al-Hamid ibn Turk and the Algebra of his Time* (Ankara, Turk Tarih Kurumu Basimenti, 1962)

Scott, J. F., *A History of Mathematics: From Antiquity to the Beginning of the Nineteenth Century* (London, Taylor and Francis, 1969)

Sedgwick, W. T., and Tyler, H. W., *Short History of Science* (New York, The Macmillan Company, 1925)

Shapley, Harlow, Rapport, Samuel, and Wright, Helen (eds.), *The New Treasury of Science* (New York, Harper and Row Publishers, 1963)

Singer, Charles, *A Short History of Scientific Ideas to 1900* (London, Oxford University Press, 1968)

Smith, David Eugene, *History of Mathematics* (New York, Ginn and Company, 1923), Vol. I

_____ *History of Mathematics* (New York, Ginn and Company, 1925), Vol. II

_____ *Number Story of Long Ago* (Washington, D.C., The National Council of Teachers of Mathematics, 1962)

_____ and Karpinski, Louis Charles, *The Hindu-Arabic Numerals* Boston, Ginn and Company, Publishers, 1911)

Somerville, D. M. Y., *The Elements of Non-Euclidean Geometry* (New York, Dover Publications, 1958)

Struik, Dirk J., *A Concise History of Mathematics* (New York, Dover Publications, 1948), Vol. I

Sullivan, J. W. N., *The History of Mathematics in Europe* (London, Oxford University Press, 1925)

Taton, Rene, *History of Science: Ancient and Medieval Science from the Beginnings to 1450* (New York, Basic Books, 1963), Vol. I

_____ *History of Science: The Beginnings of Modern Science* (New York, Basic Books, 1964), Vol. II

Taton, Rene, *History of Science: Science in the Nineteenth Century* (New York, Basic Books, 1965), Vol. III

Thorndike, Lynn, *A Short History of Civilization* (New York, F. S. Crofts and Company, 1930)

Towner, R. H., *The Philosophy of Civilization* (New York, G. P. Putnam and Sons, 1923)

Toynbee, Arnold Joseph, *A Study of History* (London, Oxford University Press, 1960), Vols. I-II

——— *A Study of History* (London, Oxford University Press, 1939), Vol. III

——— *Civilization on Trial* (New York, Oxford University Press, 1948)

Tritton, A. S., *Islam: Beliefs and Practices* (London, Hutchinson's University Library, 1951)

Van Wagenen, Theodore F., *Beacon Lights of Science* (New York, Thomas Y. Crowell Company, 1924)

Venable, Charles S., *Elements of Geometry* (New York, University Publishing Company, 1875)

Vernoeven, F. R. J., *Islam* (New York, St. Martin's Press, 1962)

Von Grunebaum, Gustave Edmund, *Islam: Essays in the Nature and Growth of a Cultural Tradition* (London, Routledge and Kegan Paul, 1961)

——— *Medieval Islam: A Study in Cultural Orientation* (Chicago, The University of Chicago Press, 1947)

——— *Modern Islam: The Search for Cultural Identity* (Berkeley and Los Angeles, University of California Press, 1962)

Wilson, Samuel Graham, *Modern Movements Among Moslems* (New York, Fleming H. Revell Company, 1916)

Yeldham, Florence A., *The Story of Reckoning in the Middle Ages* (London, George G. Harrap and Company, 1926)

Zurig, Gustantin, *Fi Ma'rakati al-Hadarah* (Beirut, Dar al-Galam Li lmlayin, 1963)

Zwemer, Samuel M., *Islam* (New York, Student Volunteer Movement for Foreign Missions, 1907)

B. Periodicals

Abbud, Faud, 'The Planetary Theory of Ibn al-Shatir: Reduction of the Geometric Models to Numerical Tables,' *Isis*, LIII (December 1962), 492

Archibald, Raymond Clare, 'Hindu, Arabic, and Persian Mathematics –600 to 1200,' *American Mathematics Monthly*, LVI (January 1949), 30

Boyer, C. B., 'Zero: The Symbol, The Concept, The Number,' *National Mathematics Magazine*, XVIII (May 1944), 323-30

Cajori, Florian, 'The Controversy on the Origin of Our Numerals,' *Scientific Monthly*, IX (November 1919), 459-63

_____ 'What Great Men Say About Mathematics,' *The Texas Mathematics Teacher's Bulletin*, III (February 1918), 41

Carnahan, Walter H., 'Geometric Solutions of Quadratic Equations,' *School Science and Mathematics*, XLVII (November 1947), 689-90

_____ 'History of Algebra,' *School Science and Mathematics*, XLVI (January 1946), 101

Eisenhower, President Dwight D., 'Address to the United Nations General Assembly,' *Arab World* (January/February 1959), 2

Gandz, Solomon, 'Arabic Numerals,' *American Mathematical Monthly*, XXXIII (January 1926), 261

_____ 'The Algebra of Inheritance,' *Osiris*, V (March 1938), 324

_____ 'The Origin of the Term Algebra,' *American Mathematical Monthly*, XXXIII (May 1926), 437

_____ 'The Sources of al-Khwarizmi's Algebra,' *Osiris*, I (21 January 1936), 264

Goldstein, R. L., 'The Arabic Numerals, Numbers, and the Definition of Counting,' *Mathematical Gazette*, XI (May 1956), 129

Jones, Philip S., ' "Large" Roman Numerals,' *The Mathematics Teacher*, XLVII (March 1954), 47

Kennedy, E. S., and Roberts, Victor, 'The Planetary Theory of ibn al-Shatir,' *Isis*, L (September 1959), 233

Khatchadourian, Haig, and Rescher, Nicholas, 'Al-Kindi's Epistle on the Concentric Structure of the Universe,' *Isis*, LVI (Summer 1965), 190-5

Kokomoor, F. W., 'The Status of Mathematics in India and Arabia During the "Dark Ages" of Europe,' *The Mathematics Teacher*,

XXIX (January 1936), 229

Landau, Rom, 'Arabist on the Cultural Heritage of the Arab World,' *The Arab World* (September-October 1960), 13

Langer, R. E., 'Euclid's Elements,' *School Science and Mathematics*, XXXIV (April 1935), 422-3

Maheu, René, 'The Development of Education in the Arab Countries,' *UNESCO Chronicle*, VI (April 1960), 137

Rankin, W. W., 'The Cultural Value of Mathematics,' *The Mathematics Teacher*, XXII (April 1929), 215

Reves, George E., 'Outline of the History of Algebra,' *School Science and Mathematics*, III (January 1952), 63

——— 'Outline of the History of Trigonometry,' *School Science and Mathematics*, LIII (February 1953), 141

Roberts, Victor, 'The Solar and Lunar Theory of ibn al-Shatir, A Pre-Copernican Model,' *Isis*, XLVIII (December 1957), 428

Said, Abdul Salam, 'We Remember that Western Arithmetic and Algebra Owe Much to Arabic Mathematicians,' *Arab World* (February 1959), 5

Sarton, George, 'The New Humanism,' *Isis*, VI (October 1924), 28

Sayili, Aydin, 'Thabit ibn Kurra's Generalization of the Pythagorean Theorem,' *Isis*, LI (March 1960), 35-6

Shahin, Nagula, 'Al-Daw'u al-Mustagtabu wa al-Tswiru al-Mghari al-Mulwan,' *Gafilh Azzit* (March/April 1972), 7-8

Slaughter, H. E., 'The Evaluation of Numbers – An Historical Drama in Two Acts,' *The Mathematics Teacher*, XXI (October 1928), 307-8

'Zero 1 2 3 4 5 6 7 8 9, Key to Numbers,' *Aramco World* (November 1961) 14

C Unpublished Materials

Haden. Marie, 'A History of Our Numerals and Decimal System of Numeration,' Unpublished Master's Thesis, George Peabody College for Teachers, Nashville, Tennessee, 1931

D. Encyclopaedias and Dictionaries

Funk, Isaac, Thomas, Calvin, and Vizetelly, Frank H. (supervisors), 'Algebra,' *New Standard Dictionary of the English Language*

(New York, Funk and Wagnalls Company)

Gibb, H. A. R., Kramers, J. H., Levi-Provencal, E., and Schacht, J. (eds.), 'Al-Battani,' *The Encyclopaedia of Islam* (London, Luzac and Company, 1960), Vol. I

James, Glenn, and James, Robert C. (eds.), 'Algebra,' *Mathematics Dictionary* (New York, D. Van Nostrand Company, 1963)

Houtsma, M. Th., Arnold, T. W., Basset, R., and Hartman, R. (eds.), 'Al-Khwarizmi,' *The Encyclopaedia of Islam* (London, Luzac and Company, 1913), Vol. I

Houtsma, M. Th., Wensiock, A. J., Arnold, T. W., Heffening, W., and Levi-Provencal, E. (eds.), *The Encyclopaedia of Islam* (London, Luzac and Company, 1927), Vol. II

Ronart, Stephan, and Ronart, Nandy, 'Al-Karkhi,' *Concise Encyclopaedia of Arabic Civilization: The Arab East* (New York, Frederick A. Praeger, 1960)

The Faculties of the University of Chicago (ed. advisors), 'Trigonometry,' *Encyclopaedia Britannica* (Chicago, Encyclopaedia Britannica, 1969), Vol. 22

E. Manuscripts

Cambridge, England, University Library, Arabic MSS, 1075, fol. 006.55

London, England, Indian Office Library, Arabic MSS, 744, fol. 1b-2a

London, England, Indian Office Library, Arabic MSS, 748, fol. 11b-12a

London, England, Indian Office Library, Arabic MSS, 748, fol. 35b-36a

London, England, Indian Office Library, Arabic MSS, 748, fol. 58b-59a

London, England, Indian Office Library, Arabic MSS, 748, fol. 83b

London, England, Indian Office Library, Arabic MSS, 749, fol. 16b-17a

London, England, Indian Office Library, Arabic MSS, 749, fol. 20b-21a

London, England, Indian Office Library, Arabic MSS, 749, fol. 57b-58a

London, England, Indian Office Library, Arabic MSS, 750, fol.

41b-42a
London, England, Indian Office Library, Arabic MSS, 757, fol. 4b-5a
London, England, Indian Office Library, Arabic MSS, 758, fol. 8b-9a
London, England, Indian Office Library, Arabic MSS, 759, fol. 45b-46a
London, England, Indian Office Library, Arabic MSS, 760, fol. 29b-30a
London, England, Indian Office Library, Arabic MSS, 767, fol. 198b-199a
London, England, Indian Office Library, Arabic MSS, 771, fol. 20v-33
London, England, Indian Office Library, Arabic MSS, 771, fol. 33b-34a
London, England, Indian Office Library, Arabic MSS, 772, fol. 17b-18a
Oxford, England, Bodleian Library, Arabic MSS, 119, fol. ff. 49r-54r
Oxford, England, Bodleian Library, Marsh MSS, 489, fol. 145r-166r
Oxford, England, Bodleian Library, Marsh MSS, 640, fol. f. 102
Paris, France, Sorbonne University, Arabic MSS, 2544, fol. Gall, SI. 374

INDEX

Abd-Al-Malik 22
Abd-Al-Rahman 24
Abdul Rihan Mohammed ibn Ahmed
Al-Biruni 10-11, 73
Abu Bakr 20-21
Ali, ibn Abi Talib 21
agriculture 12
Ahmes 83
algebra 9, 49-65, 89, 95; definition of 49-50; origin of the term 50; interrelation of, with geometry 82, 93, 100
algorithms 33
al-kasr 42-3
amicable numbers 43-4, 95
Apollonius 13, 44, 83, 86
Arabia 96-101; history of 19-27; the Abbasid Caliphate 23-4; the Caliphate 20-21; the Umayyad Caliphate 22-3
Arabic 98
Arabic numerals 33-7
Archimedes 13, 44, 86, 89
Aristotle 10, 14, 75, 85
arithmetic 9, 31-47; operations 39-42
astrolabe 76
astrology 71
astronomy 12, 13, 25, 89, 94, 96; trigonometry and 67, 70-75, 100

Babylonians 83, 94
Bacon, Roger 84
Baghdad 10
Banks, J. Houston 34
Battani, Mohammed ibn Jabir ibn Sinan Abu Abdullah Al- 13, 70-73
Bell, Eric T. 13
Boyer, Lee E. 36
Briffault, Robert 15
Byng, Edward J. 76

Cajori, Florian 88
Cavalieri, Francesco B. 38
celestial navigation 76
chemistry 12
Chuquet, Nicholas 38
circles 88
compass 76
Conant, Levi, L. 45
Copernicus 69, 74
cosine 69, 71
cotangents 13, 71, 75
cubic equation 59-60, 76

decimal system 36
Diophantus 52, 60, 93, 95
division 39, 41, 42; algebraic 59
double false position 60-62

Eddin, Beha 50
Egypt 83
Eisenhower, Dwight D. 98
equations 49-50; first degree 50, 55; higher than second degree 13; linear and quadratic equations 51, 55-9, 60; real root of 60; second degree 50
Euclid 10, 14, 44, 76; geometry 81-9 *passim*
Euler, Leonhard 38
Europe 97-8; adoption of zero 37
Eutocious 44
Eves, Howard 61, 85

Fermat, P. 59-60
Fibonacci, Leonardo 37, 39
Fink, Karl 59
fractions 42-3; decimal 43

Galen 75, 85
Galileo 67, 95
Gandz, Solomon 50, 52
Genghis Khan 25-6
geography 12
geometry: algebraic 55-9, 82, 89, 93, 100; definition of 82; origin of 82-3

Gerhard of Cremona 52
Gershon, Levi Ben 72
Goodstein, R.L. 45
Greece, ancient 9, 11-12, 14-15, 83, 85-6; translations from Greek 10, 13, 15, 25, 44, 81, 89

Haitharn, ibn Abu-'Ali al Hasan ibn Al-Hasan Al- 67, 75, 83-6
Hajjaj, ibn Yusuf Al- 81, 88
Hakam, Al- 24-5
Herigone, Pierre 38
Herodotus 83
Hindu mathematics 10, 34, 37, 49
Hitti, P.K. 22, 32
House of Wisdom (Bait Al-hikma) 10
Howe, George 68

integers 59
Islam *see* Arabia

Joseph, ibn 'Acquin 84

Khaldun, ibn Abu Zaid Abdel Rahman 43, 49, 81-2
Karkhi, Abu Bekr Mohammed ibn Al-Hosain Al- 13, 32, 45, 53, 60, 63
Karpinski, Louis C. 43, 51
Kepler, Johann 84
Khwarizmi, Mohammed ibn Musa Al- 13, 23-4, 32-3, 49, 52-63 *passim,* 74-5, 88, 89
Kindi, Abu-Yusef Ya'quib ibn Ishaq Al- 31-2, 87-8
Kokomoor, F.W. 11-12
Koran, the 9, 19
Kuhi, Waijan ibn Rustan Abu Sahl Al- 13

Lambert's quadrangle 86
Lilavati 42-3
linear equations 51, 55-9, 60
logic 14
long division 41-2

Maheu, René 98
Ma'mum, Al- 10, 23, 50, 74, 81, 88
Mansur, Al- 81
medicine 12
Metiers, Adrian 38
Miller, George A. 11
Mohammad Khan 51
Mohammed 9, 10, 19, 94
Mongols, the 97
moon, the 84
multiplication 39; algebraic 59; lattice method 39-40
Muslim State 10
Musta 'sim, Al- 23

Napier, John 96
natural numbers, sum of 45
navigation 76
Newton, Sir Isaac 67
numerals and numbers 95; amicable 43-4, 95; Arabic 33-7; natural 45; rational 60; Roman 33-4; theory of 13

Omar, ibn Al-Khattab 21
optics 67; geometrical 75, 83-4, 94
Othman, ibn Affin 21
Ottoman Turks 26-7, 97

Pacisli, Luca 38
parallelograms 88
Peckham, John 84
Peurbach, Georg von 69
pharmacology 12
physics 89
Pitiscus 68
Planudes, Maximus 37
Ptolemy 10, 44, 69, 72-3, 81, 84, 86, 89
Pythagoras 14, 89

quadratic equations 51, 55-9, 60

radical axis, theorem of the 84
Rashid, Harun Al- 23, 81, 88
rational numbers 60
Reeve, William, D. 82
Regiomontanus 69, 72
religion 9, 15-16, 94, 99
Renaissance, the 94
Robert of Chester 52
Roman numerals 33-4
roots 52-9

Said, ad-Dimishqui 88
Said, Hakim Mohammed 86

Sarton, George 11, 15, 50, 52, 94
science 94-5
Shatir, ibn ala al-Din Ali ibn Ibrahim Ali- 73-4
Sicily, Muslims in 25
sines 69-70, 71, 75
Smith, David E. 49-50, 51
Snell, Willebrord 75
Spain, Muslims in 24-5
spherical geometry 87-8
spherical triangles 71, 76
square root 53
Struik, Dirk J. 14-15
successive approximations 95
Sufyan, Abu ibn 21-3
surveying 89

Taki Ed Din al-Hilali 73
tangent 69, 71
Thabit, ibn Qurra 13, 44, 59, 86-7, 89
Thales of Miletas 83
trigonometry 13, 67-79, 88, 95-6, 100; definition of 68; origin of 68-70
Turks, the 26-7, 97

Vinci, Leonardo da 67, 84

Walid, Al- 22-3
Wathiq, Al- 23
Western culture 97-8
Wiedmann, E. 50-51
Witelo, 84

zero 37-9, 96